Vacation
& Leisure
Home Plans

Vacation & Leisure Home Plans

Edited by Garlinghouse

THE GARLINGHOUSE COMPANY
Topeka, Kansas

VACATION & LEISURE HOME PLANS

A Garlinghouse Company publication.

ISBN 0-938708-05-8

Printed in the United States of America.

Library of Congress catalog card number 82-084444.

Simple, easy to build passive plan

No. 26950—The design of this plan allows for excellent air circulation between downstairs living areas and second level rooms. Ceilings rise two stories in the family, dining, and living areas and are open over balcony railings to upstairs bedrooms. The living room features deck access and wood burning fireplace. The conveniently arranged kitchen is highlighted by a generous pantry. A unique half round deck lies off the entry way and another roofed deck connects the garage and house. The shape of the double garage allows it to be entered from any of three directions that the home builder might choose.

First Floor—1,090 sq. ft.
Second Floor—580 sq. ft.
Garage—484 sq. ft.

*For price and order information
see pages 108-109*

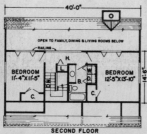

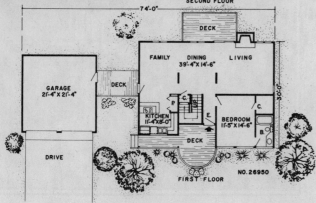

Vacation & Leisure Home Plans

1

A home
with distinction

No. 26952—A wood shake roof, vertical cedar siding and stone trim mark this home with distinction. The centrally located great room is the hub of activities. A sunken area is found here with wet bar and cabinets on one side and fireplace on adjacent side. Opposite is a full length planter with skylights above. Planters just outside give a continuity between indoor and outdoor spaces. Cathedral ceilings continue from the great room through the master bedroom and bath and side deck. The side deck is handy to two smaller bedrooms and master bath. A large deck area, rimmed by an abundance of glass and planters, is found across the rear. A utility area is conveniently placed between the double garage and kitchen area. The large, well designed kitchen boasts an eating bar for quick meals.

First Floor—2,484 sq. ft.
Garage—418 sq. ft.
Sauna & pool bath—84 sq. ft.
Decks—635 sq. ft.

*For price and order information
see pages 108-109*

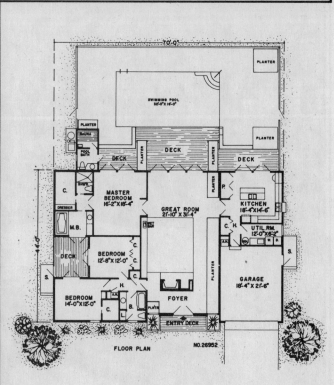

FLOOR PLAN   NO. 26952

3

Ornate windows featured

No. 26782—Charm is found inside and out in this modified victorian style solar design. Cedar shakes, narrow horizontal siding, an ornate window arrangement and varying roof angles add to the exterior appeal. Inside, an air lock entry, heat circulating fireplace in the great room, and two story greenhouse with concrete and brick floor for heat storage, combine to produce an energy efficient design. A spacious formal dining, combination great room, highlight the lower level. The kitchen lies snuggled between a multi-purpose room and casual eating area. A master bedroom suite, additional bath, bedroom and bedroom/den occupy the second level. Openings and windows between second level rooms and upper portions of the foyer, great room, and greenhouse enhance air circulation. An extra large double garage completes the design.

Upper level—1,078 sq. ft.
Lower level—1,108 sq. ft.
Garage—618 sq. ft.
Deck—240 sq. ft.
Greenhouse—196 sq. ft.

*For price and order information
see pages 108-109*

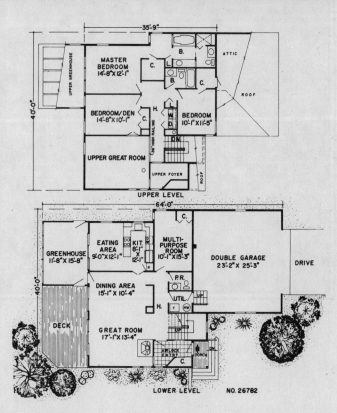

UPPER LEVEL

- 35'-9"
- 40'-0"
- UPPER GREENHOUSE
- MASTER BEDROOM 14'-8"x 12'-1"
- B.
- C.
- ATTIC
- B.
- C.
- ROOF
- BEDROOM/DEN 14'-8"x 10'-1"
- H.
- W D
- BEDROOM 10'-1"x 11'-5"
- C.
- UPPER GREAT ROOM
- 36" HIGH RAILING
- DN.
- UPPER FOYER
- ROOF

LOWER LEVEL   NO. 26782

- 64'-0"
- 40'-0"
- C.
- GREENHOUSE 11'-8"x 15'-8"
- EATING AREA 9'-0"x 12'-1"
- KIT. 8'-1" X 12'-1"
- MULTI-PURPOSE ROOM 10'-1"x 15'-3"
- DOUBLE GARAGE 23'-2" x 25'-3"
- DRIVE
- DINING AREA 15'-1" X 10'-4"
- P.R.
- H.
- UTIL.
- UP
- DECK
- GREAT ROOM 17'-1"x 13'-4"
- UP
- AIRLOCK ENTRY
- PORCH
- DN.
- C.

5

Small, gable roof design

No. 10400—Living spaces flow freely in the floor plan of this economical home. The master bedroom suite is privately separated from two smaller bedrooms and a bath by the more active kitchen and family room areas. A wood burning fireplace adds warmth to a more formal living room. Plenty of insulation is called for in the building plans.

First Floor—1,184 sq. ft.

For price and order information see pages 108-109

FLOOR PLAN

NO. 10400

BEDROOM
10'-4" X 9'-8"

FAMILY
ROOM
10'-6" X 13'-0"

LIVING ROOM
17'-4" X 14'-4"

BEDROOM
10'-4" X 9'-6"

KITCHEN
8'-6" X 14'-0"

ENTRY

MASTER
BEDROOM
12'-4" X 10'-4"

44'-0"

28'-0"

7

Passive solar leisure home

No. 10390—Passive solar features and leisure oriented living spaces create the focus of this home. Sunwall panels, an airlock entry, greenhouse, two solar furnaces and a wood burning, circulating fireplace reduce energy consumption needs. Acting as a buffer to winter winds is a northerly placed double garage. A shop area, hobby room, large living room and open kitchen/eating bar/family room with adjoining deck encourage relaxed activities. The kitchen sports its own window greenhouse. Two bedrooms are shown on the upper level with several options on the lower level for additional bedrooms and a bath.

First Floor—1,293 sq. ft.
Basement—1,232 sq. ft.
Garage—576 sq. ft.
Greenhouse—192 sq. ft.
Deck—240 sq. ft.

For price and order information see pages 108-109

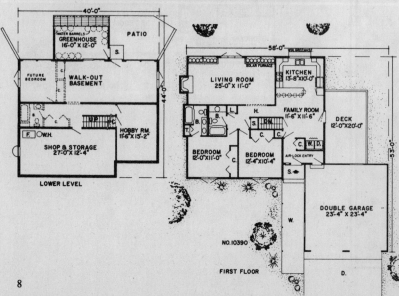

Three levels of living space

No. 10396—This passive solar design is suitable for vacation or year round living. The rear or southern elevation of the home is highlighted by an abundance of decks and glass. A minimum of windows are found on the north, east and west sides. On the basement level are found large shop, storage, and recreation areas plus a bedroom. Unique aspects of the first level living room are its location up two steps from the rest of the first floor, two stories of glass on its southern wall and its openness with a railing to the hall and stairway adjacent. An angled wall lends character to the kitchen/dining area. The master suite occupies the entire second level with its own bath, dressing area, walk in closet, storage nook and private deck.

First Floor—886 sq. ft.
Second Floor—456 sq. ft.
Basement—886 sq. ft.

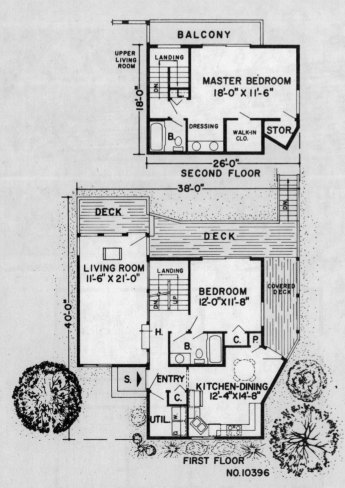

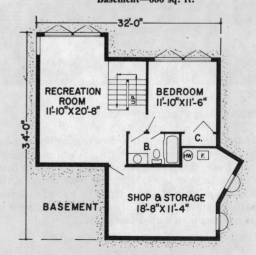

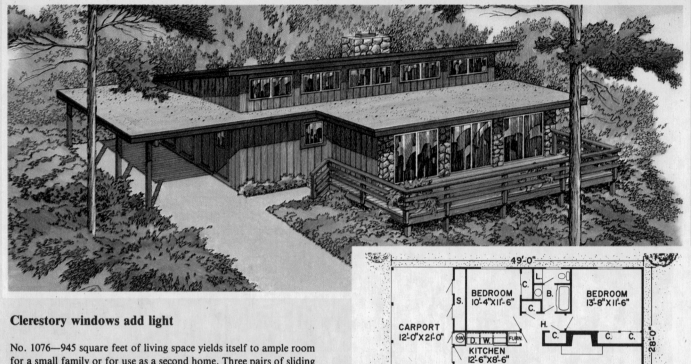

Clerestory windows add light

No. 1076—945 square feet of living space yields itself to ample room for a small family or for use as a second home. Three pairs of sliding glass doors, bordered by rock columns, give a striking exterior appearance as well as furnishing an excellent view from the living and dining rooms within. Clerestory windows rise above the living/dining room area and kitchen, illuminating them with light. Ceilings slope away to the rear of the design from the clerestory windows to a full bath, flanked by two bedrooms, creating full or partial cathedral ceilings in all rooms of the house.

Living area—945 sq. ft.
Carport—252 sq. ft.   Deck—252 sq. ft.

Comfortable cottage suits narrow lot

No. 8082—Adaptable to a 50-foot lot, this small cottage boasts an exterior of horizontal siding, brick, and shutters, as well as a cozy interior. Entry is directly into the living room, splashed with light from plentiful windows. Large enough to entertain a group of people, the living room is shut off from sleeping quarters by a door, which encourages maximum privacy and quiet. Two adequate bedrooms and full bath are set opposite an extra storage closet.

First floor—936 sq. ft.; Basement—936 sq. ft.

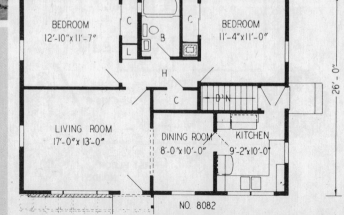

Vacation retreat or year round living

No. 1078—A long central hallway divides formal from informal areas, assuring privacy for the two bedrooms located in the rear. Also located along the central portion of the design are a utility room and neighboring bath. The furnace, water heater and washer dryer units are housed in the utility room. An open living/dining room area with exposed beams, sloping ceilings and optional fireplace occupies the design's front. Two pairs of sliding glass doors access the 411 feet of deck from this area. The house may also be entered from the carport on the right or the deck on the left.

First Floor—1,024 sq. ft.
Carport & Storage—387 sq. ft.   Deck—411 sq. ft.

Design features six sides

No. 1074—Simple lines flow from this six-sided design, affordably scaled, but sizable enough for a growing family. Active living areas are snuggled centrally between two quieter bedroom and bath areas in the floor plan. A small hallway, accessing two bedrooms and a full bath on the right side, may be completely shut off from the living room, providing seclusion. Another bath lies behind a third bedroom on the left side, complete with washer/dryer facilities and close enough to a stoop and rear entrance to serve as a mud room.

First Floor—1,040 sq. ft.   Storage—44 sq. ft.
Deck—258 sq. ft.   Carport—230 sq. ft.

11

Small apartment in full basement

No. 10370—This floor plan lends itself to a wide range of possibilities. The basement includes a kitchen, dining room, fireplaced family room, bath and bedroom, excellent for use by your own family or as a separate apartment. Sliding glass doors access a patio from both the family room and bedroom. A large shop on this level provides your do-it-yourselfer with plenty of workroom for any project. Sloping ceilings with exposed false beams and fireplaces adorn both the large living and dining rooms on the upper level.

**Upper Level—1,353 sq. ft. Lower Level—1,353 sq. ft.
Garage—576 sq. ft.**

Design suited for passive and active solar use

No. 26920—Suited well for a site with a south facing slope, this dwelling features thick insulation, berming, an air lock entry and passive solar heat gain to minimize energy needs. Quiet and active areas are well separated in the floor plan. Traffic channels have been designed to avoid intruding through rooms. Living room highlights are a dramatic high ceiling and wood burning fireplace. Wood treatment of the exterior, rough textured roof and the design's simple shape give it a rugged appearance.

**Main Floor—840 sq. ft. Lower Level— 840 sq. ft.
Garage—574 sq. ft.**

Design combines laundry, bath

No. 10268—A large, closeted laundry and half bath combination show the emphasis on space and function in this attractive two story. The living room extends 20 feet to feature a fireplace, and the dining room opens to the patio via sliding glass doors. Zoned for sleeping, the upstairs groups three sizable bedrooms and two full baths, adding plentiful closet space.

First floor—882 sq. ft., Second floor—875 sq. ft.,
Covered patio—600 sq. ft.,
Garage—576 sq., ft., Basement—882 sq. ft.

Eye-catching angle marks vacation plan

No. 10224—Fronted by a redwood deck, an exceptional exterior blends with a practical floor plan in this refreshing leisure home plan. Entry and lounge is equipped with closets and bordered by bedroom on one side and living area on the other. Corridor kitchen and dining area open to a handy porch. Upstairs, the bunk room offers ample sleeping space for guests and, with the hobby room, boasts an enjoyable second story deck.

First floor—1,012 sq. ft., Second floor—406 sq. ft.

13

Lakeshore home spells luxury

No. 10138—Take the luxurious features of a second floor sun deck, ground-level patio, and spacious living and family rooms; combine with four full size bedrooms; add to this two and one-half baths in a unique and practical arrangement, a ground-floor utility room and a well-grouped kitchen.

Upper level—1,196 sq. ft., Lower level—1,196 sq. ft.

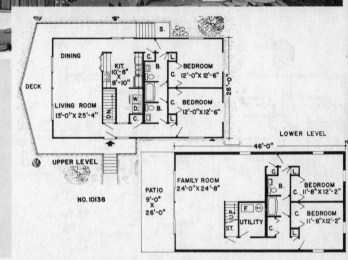

Complements natural surroundings

No. 10134—Simplicity of external design aids in blending this vacation home with the area in which it is located. A large boat storage area has its own entrance to the outside and could be modified into a recreation room or work room. Large, well-sheltered porches provide a view both in front and back.

Upper level—840 sq. ft., Lower level—840 sq. ft.

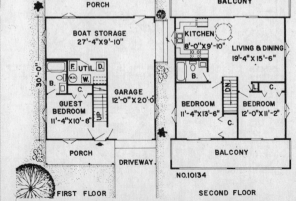

Breakfast bar sets off dining area

No. 10054—Devised for vacation living, this Chalet beach home features several worksaving ideas, including a breakfast bar which divides living room and kitchen. The ample living-dining room spills out onto the attractive 24-foot deck. Four closeted bedrooms include two upstairs, favored with balconies and reached by a spiral staircase off the living room. The home is built on treated pilings but might also be constructed on a conventional foundation.

First floor—768 sq. ft., Second floor—406 sq. ft.

A view from each room

No. 10116—Every bedroom of this attractive vacation home offers it own private porch or balcony connected by sliding glass doors. Access to the second floor is by a modern spiral staircase. Ample closet space and a large storage area are provided for. This design makes use of a carport located near the kitchen to save many steps after shopping.

**First floor—720 sq. ft., Second floor—576 sq. ft.,
Carport—213 sq. ft.**

Singular home practical to build

No. 9950—Distinctive as this home may appear, with its deck-encircled hexagonal living room, its construction will actually prove practical. Besides the living room, which exhibits exposed beams and a cathedral ceiling, the main level encompasses four bedrooms, two baths, dining room and kitchen. On the lower level, an enormous family room opens to a patio, with built-in barbecue. Another bedroom, den and bath with shower are detailed. Boat storage is also provided on this level.

First floor—1,672 sq. ft.,
Lower floor—1,672 sq. ft., Garage—484 sq. ft.

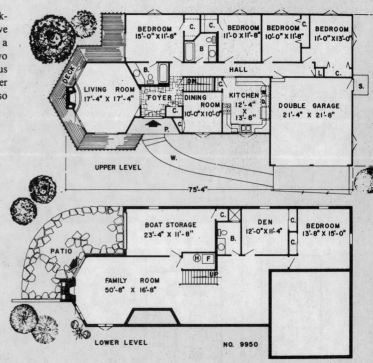

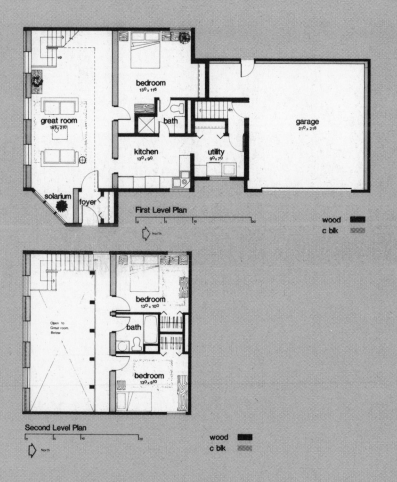

First Level Plan

North

wood

c blk

Second Level Plan

North

wood

c blk

Passive solar for all surroundings

No. 28016—This spacious, contemporary-styled home has 1,900 square feet of living space with 3 bedrooms, 2 baths, a 2-car garage and a Great Room with a two-story cathedral ceiling and clerestory windows. The house features 2x6 exterior wall, R-27 wall insulation and R-60 ceiling insulation, an air-lock entry, double-and triple-glazed windows, insulating curtains and a thermal storage wall.

First Floor—1,057 sq. ft.
Second Floor—851 sq. ft.
Basement—976 s. ft.
Garage—542 sq. ft.

Old fashioned charm

No. 21124—An old fashioned, homespun flavor
has been created using lattice work, horizontal and
vertical placement of wood siding, and full length
front and rear porches with turned wood columns
and wood railings. The floor plan features an open
living room, dining room and kitchen. A master
suite finishes the first level. An additional
bedroom and full bath are located upstairs. Here,
also, is found a large bonus room which could
serve a variety of family needs. Or it can be deleted
altogether by adding a second floor balcony overlooking the living room below and allowing the living room ceilings to spaciously rise two full stories.
Wood floors throughout the design add a final bit
of country to the plan.

First Floor—835 sq. ft.
Second Floor—817 sq. ft.

Materials list not available for this plan.

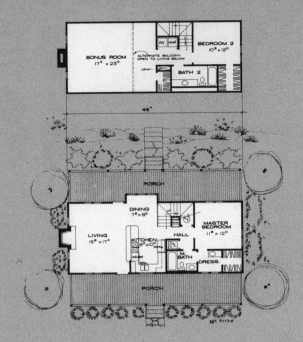

Elegant living

No. 21130—Architectural geometry is heightened
throughout this beautiful home by touches of elegance. A generous entry foyer separates formal
and informal areas. To the foyer's left, you enter
either the living room or dining room, separated
from each other by a built-in bookcase hutch. The
dining room looks out onto and accesses a foun-
tained garden. Moving on around the design, a triangular cutting counter, range/oven island highlights the kitchen. Dividing the kitchen and family
room is an eating bar. Featured in the family room
are a bar area and terrace entry. The central hub of
this portion of the plan provides a pantry for the
kitchen and fireplaces for both the living and family rooms. Excluding one small living room space,
full length windows line all exterior walls.

First Floor—3,014 sq. ft.
Garage—877 sq. ft.
Storage—120 sq. ft.

Materials list not available for this plan.

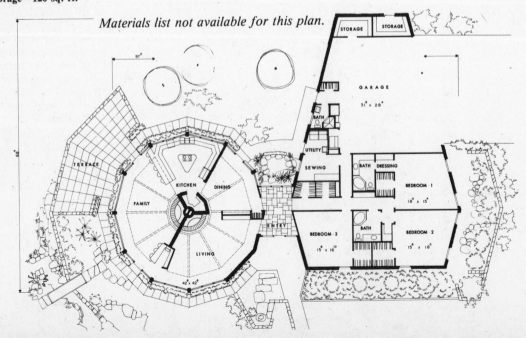

aterials list not available for this plan.

A contemporary with character

No. 21128—Variety in exterior materials and an abundance of decks, roof levels, glass, exposed 4 x 16 beams and outcroppings blend to give exterior character to this contemporary. The floor plan continues the uniqueness. The entry leads you through a graceful arch to the formal dining room. To the entry's left, separated by railings and down two steps, is the living room. It sports a fireplace and, like the dining room, two stories of glass windows. Vaulted ceilings are found in these two rooms plus the entry. A deck on this level can be approached from both living room and informal breakfast area. The kitchen, a full bath and versatile sewing room finish the first level floor plan. Stairs, open on the left to the entry and living room, sweep to the second level. A secluded study forms a focal point here. Walls in this room are unbroken by windows and are perfect for bookshelves. Shutters open to the living room below and aid air circulation. Double doors add privacy. A deck is accessible from the study or master bedroom. The master bedroom with nearby bath and large closet, features a bay window. Another full bath, two bedrooms and washer/dryer facilities have been added to complete the second level. Spaces for the water heater and A/C unit are provided on the lower level. A double carport ends the design.

First Floor—1,179 sq. ft.
Second Floor—1,022 sq. ft.
Carport—328 sq. ft.

*For price and order information
see pages 108-109*

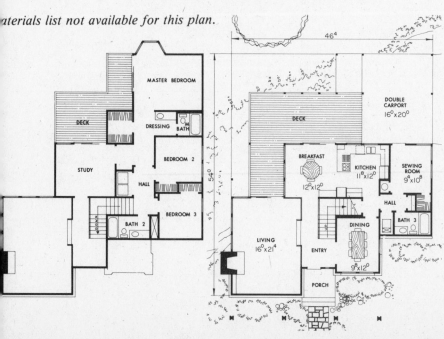

19

Relaxed and economical

No. 21126—Well suited for the economy minded small family or as a second home is this design. To the left of a large front entry lies the living room. Here you find access to a deck, a fireplace and a cathedral ceiling with exposed beams. The living room flows through an eating bar to the kitchen/dining area beyond. The dining room also accesses the deck. To the right of the entry are two bedrooms and a full bath. Sliding glass doors and full length windows cloak the entire width of the rear of the house on this level. A touch of elegance is provided by a stairway spiraling to the second floor loft. Clerestory windows draw in the sun and illumine this quiet, secluded room.

First Floor—1,082 sq. ft.
Loft—262 sq. ft.

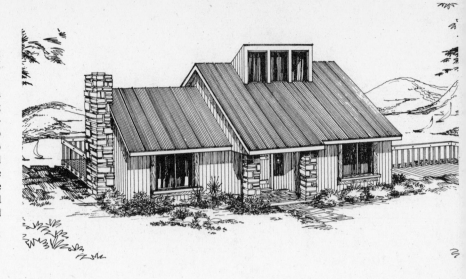

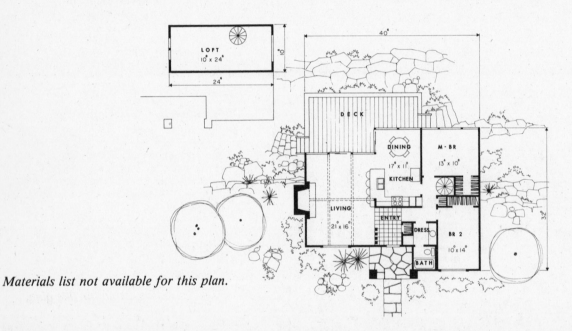

Materials list not available for this plan.

A plan with lofty ideas

Materials list not available for this plan.

No. 21120—An attractive beginning for the first floor plan is found in the centrally located foyer. Traffic is easily directed from here to all parts of the design. Down several steps lies a sunken family room, accentuated by two stories of glass at its rear, cathedral ceilings and a fireplace. The family room shares openness of design with the adjacent kitchen and dining room. Stairs spiral from the foyer to a second level loft. The loft overlooks the family room below from a full length balcony on one side and accesses a private exterior deck through glass doors on the other side. A master bedroom with its own bath lies secluded to the left of the foyer. An additional bath and washer/dryer facilities are located off the hall to the master suite. A carport, 49 square feet of storage at its rear and a covered terrace, accessed from the kitchen and family room, complete the plan.

First Floor—947 sq. ft.
Second Floor—232 sq. ft.
Carport—346 sq. ft.
Storage—49 sq. ft.

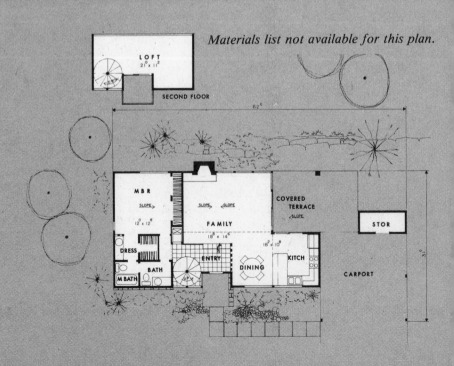

Energy efficient and compact

No. 21122—This compact dwelling is simple but organized and efficient. The open floor plan calls for a kitchen, dining room, and living room area unbroken by walls. All ceilings in the design slope to a peak above this open area. Fixed glass has been placed in every nook and cranny left by the sloping ceilings and, together with abundant sun catching windows, adds an energy saving feature to the already advantageous excellent air circulation provided for by the design's openness. The massive stone fireplace and wood storage bin attract your eye in the living room. Two bedrooms, a full bath and carport are located to the right of the plan. A storage area which also houses the washer/dryer facilities and water heater can be accessed from the carport. A large deck lies to the rear of the home.

First Floor—1,062 sq. ft.
Carport—242 sq. ft.
Storage—77 sq. ft.

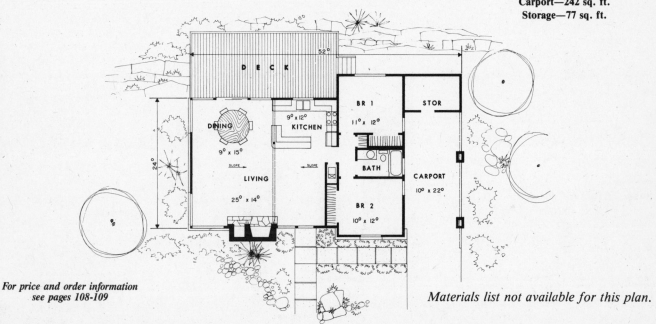

*For price and order information
see pages 108-109*

Materials list not available for this plan.

Living areas arranged to savor view

No. 10078—Encircled by deck and balcony and open to the outdoors via three sets of sliding glass doors, the living room in this appealing home is set to enjoy the scenery. Entry is into the lower level foyer, with bedroom at left ideally located to serve as a home office. Full bath and family room border a utility room with access to the carport-patio with built-in barbecue grill. Two bedrooms upstairs share a compartmented bath, and the living room enjoys a wood-burning fireplace.

Upper level—1,116 sq. ft.,
Lower level—724 sq. ft.,
Carport and patio—526 sq. ft.

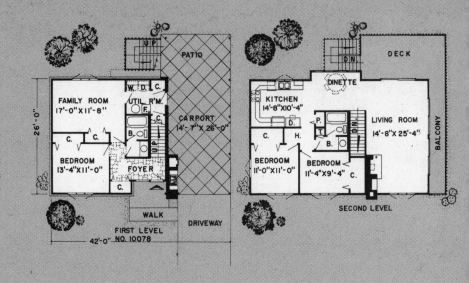

Plan takes advantage of view

No. 10062—This leisure home takes full advantage of frontage of views and outdoor living. Designed for a sloping lot, it will fit equally well on a lakeshore or mountain site. The main floor contains two bedrooms and a large bathroom which includes space for the laundry equipment. Open planning is used in the kitchen-family room area which adds spaciousness. The lower level, reached via a spiral stairway, features a family room with a wood-burning fireplace.

Floor plan—816 sq. ft., Lower level—816 sq. ft.

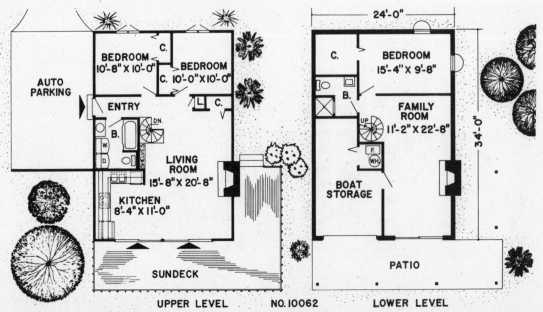

Glass doors provide view, access

No. 10066—Sporting an exterior that is dramatic and heavily glassed, this two bedroom vacation home is styled for enjoyment of scenery and access to outdoor living. Below, the covered patio holds a masonry barbecue grill and adjoins a boat storage area with built-in workbench. On the main level, sliding glass doors open firelit living room, kitchen, and front bedroom to the deck. The extra large bath outlines a laundry niche and towel closet, and both bedrooms are closeted and spacious.

Upper level—898 sq. ft.,
Lower level—336 sq. ft., Carport—281 sq. ft.

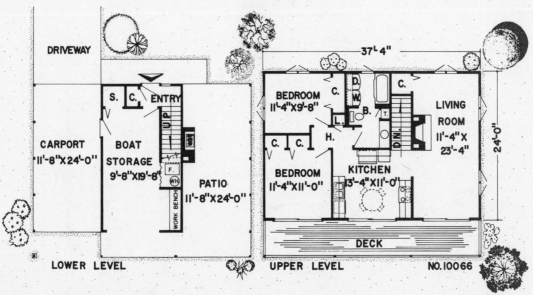

Window expanses blend home, nature

No. 10086—Expressing drama in design, this year-round leisure home offers bright, expansive living and sleeping areas in a setting in touch with nature. The extensive use of glass absorbs daylight and scenery, and sliding glass doors are doubly useful in creating access to outdoor living areas.

**First floor—1,095 sq. ft.,
Second floor—830 sq. ft.**

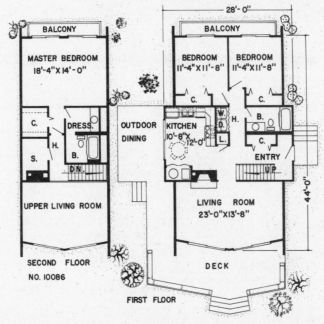

BALCONY

MASTER BEDROOM
18'-4" X 14'-0"

C.

DRESS.

H.

B.

S.

DN.

UPPER LIVING ROOM

SECOND FLOOR
NO. 10086

OUTDOOR DINING

28'-0"

BALCONY

BEDROOM
11'-4"X 11'-8"

BEDROOM
11'-4"X 11'-8"

C.

C.

W. H.
D.
L.

B.

C.

KITCHEN
10'-8" X 12'-0"

ENTRY

UP

LIVING ROOM
23'-0"X13'-8"

44'-0"

DECK

FIRST FLOOR

Contemporary plan ideal for narrow lot

No. 10034—Besides suiting the narrowest of mountain or lake sites, this contemporary design, layered with shake shingles and cedar siding, would enhance a confining urban lot as well. Adapted to all-season living, it houses an open living room, spanning one half the home and adorned with a bow window, a formal dining room separated by sliding glass doors from the front porch, a kitchen and bath on the first floor. The upstairs is comprised of a bath and three bedrooms, one of which opens onto a sun deck.

**First floor—824 sq. ft.,
Second floor—824 sq. ft., Basement—824 sq. ft.**

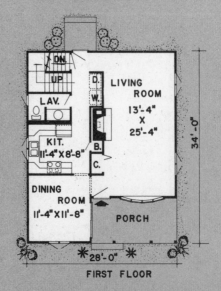

DN.
UP

LAV.

D. W.

LIVING ROOM
13'-4"
X
25'-4"

KIT.
11'-4"X8'-8"

B.

C.

DINING ROOM
11'-4"X11'-8"

PORCH

34'-0"

28'-0"

FIRST FLOOR

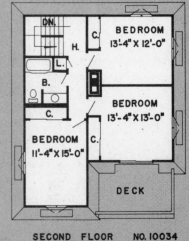

DN.

H.

C.

BEDROOM
13'-4" X 12'-0"

L.

B.

C.

BEDROOM
13'-4" X 13'-0"

BEDROOM
11'-4" X 15'-0"

C.

DECK

SECOND FLOOR   NO. 10034

Two bedrooms enjoy view, balcony

No. 10002—Two of the three bedrooms in this home open into the redwood balcony via sliding glass doors and share a potentially lovely view. Another bedroom, a large bath with towel closet, and a hall linen closet complete the upper floor. On the main level, the living room, favored with a wood-burning fireplace, borders the roomy kitchen which incorporates an informal dining area. A utility room, bath, and substantial closet space are also supplied on this level.

**First floor—624 sq. ft.,
Second floor—624 sq. ft.,
Carport and storage—432 sq. ft.**

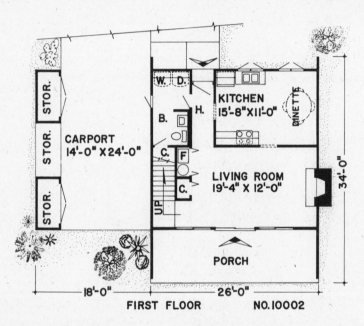

FIRST FLOOR   NO. 10002

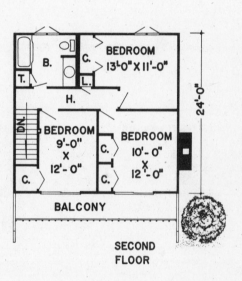

SECOND FLOOR

Two terraces featured

No. 9502—Plank and beam construction is used throughout the house for the ceilings and roof. The semi-open stairway in the entry leads to a large recreation room in the basement which has a massive cut stone fireplace. The lower terrace is more informal than the upper one and has a barbecue oven. An ornamental wood fence shields this area from the street. Any room in the house can be reached from the front entry or kitchen without going through another room.

**First floor—1,687 sq. ft.,
Basement—1,687 sq. ft.**

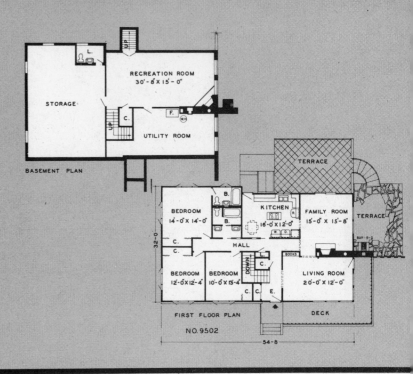

Deck encourages relaxation

No. 1072—The deck and rustic appearance of this dwelling provide an atmosphere which encourages relaxation. Through sliding glass doors, the deck accesses the living room, flowing through a dining area to the kitchen beyond. Laundry facilities and linen shelves are close at hand around the corner from the kitchen. The plan's rear is reserved for a larger master bedroom, small second bedroom, and full bath. Traffic is well handled in the design by the provision of two rear entrances and several hallways. An optional carport plan is shown.

**First Floor—868 sq. ft.
Decks—283 sq. ft.
Storage—29 sq. ft.**

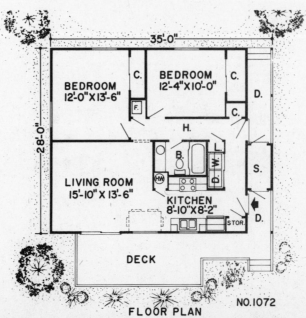

Small three bedroom

No. 1070—Suitable for a narrower lot, this well organized design provides for three bedrooms and one and a half baths in just 1064 square feet of livable space. Fixed glass above the door and a sidelight splash the front entrance with light, providing warmth and cheeriness as you enter. The living room sports a large front expanse of glass, is open to the dining area and situated away from traffic. Washer and dryer units are conveniently placed in the kitchen, which, like the baths and bedrooms, are serviced by a central hallway. Vaulted ceilings are found through the design except in bathroom areas. A rear entrance from the kitchen to the carport/storage area is provided.

First Floor—1,064 sq. ft.
Carport & Storage 288 sq. ft.

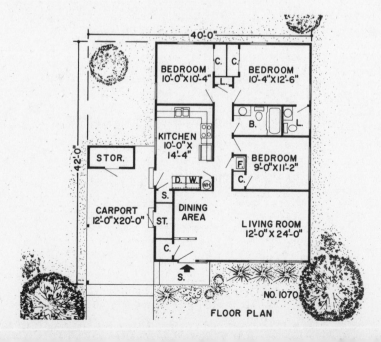

FLOOR PLAN

Vacation retreat captures light, air

No. 10196—Heavily glassed to ensure sunlight and scenery, this three bedroom refuge promises totally comfortable living all year long. The entry level spotlights a 29-ft. living room, open to the patio through sliding glass doors. Another patio annexes the dining area, and a sizable den promises space for reading, watching television, or accommodating extra guests. Nestled in the middle level are three bedrooms, one of which opens to a private deck. The large studio, an undisturbed area for quiet and privacy, tops the design.

Lower level—1,089 sq. ft.,
Middle level—652 sq. ft.,
Upper level—306 sq. ft.

For price and order information see pages 108-109

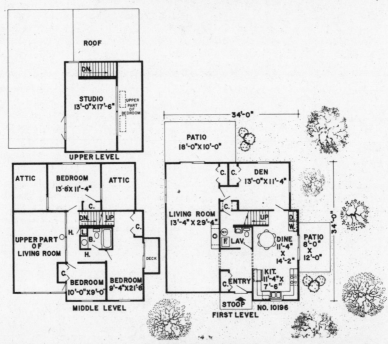

Vacation fourplex set for sports

No. 10290—Well-designed to encourage outdoor fun and indoor relaxation, this four unit leisure plan seeks a scenic spot near a winter sports area. Each of the units is indulged with a private deck that opens to a family room/kitchen. On the first floor, the units share a common living room, warmed by a wood-burning fireplace, while on the lower level, they can enjoy a sauna, television room, pool room, game room, and ski room. The plan even details a wine storage area, and balconied lofts provide sleeping space on the second level.

First floor—1,650 sq. ft.,
Second floor—980 sq. ft.,
Basement—1,650 sq. ft.

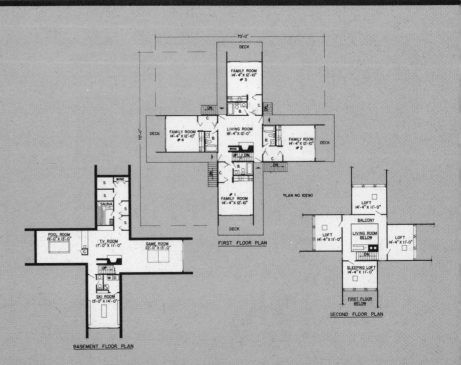

Siding, shingles adapt design to nature

No. 10218—Rustic log cabin siding and cedar shingles fuse with wooden deck and stone chimney to craft a design that belongs in and joins with natural surroundings. Furnished with two bedrooms on the lower level, the plan supplies a balconied sleeping loft that stretches 27 feet to provide sleeping space for family or guests. On the main level, the living room merits a restful fireplace and adjoins a full bath, while the compact kitchen shows dining space and sliding glass doors to the deck.

**First floor—648 sq. ft.,
Second floor—360 sq. ft.,
Basement—476 sq. ft., Garage—352 sq. ft.**

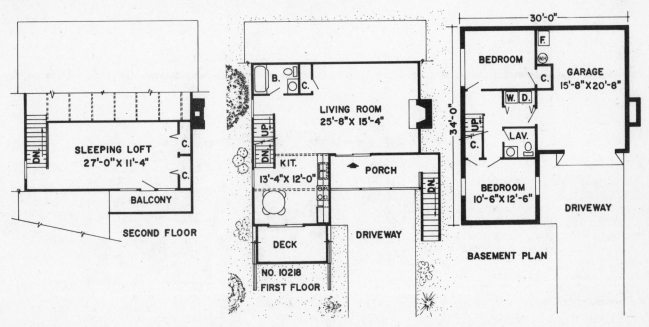

SLEEPING LOFT 27'-0" X 11'-4"

BALCONY

SECOND FLOOR

B.   C.

LIVING ROOM 25'-8" X 15'-4"

KIT. 13'-4" X 12'-0"

PORCH

DECK

NO. 10218 FIRST FLOOR

DRIVEWAY

30'-0"

F.

BEDROOM

W.H.

GARAGE 15'-8" X 20'-8"

C.

W.   D.

34'-0"

LAV.

BEDROOM 10'-6" X 12'-6"

DRIVEWAY

BASEMENT PLAN

Roofed porch skirts vacation plan

No. 10132—Stone chimney and roofed porch both embellish the exterior and indicate interior livability of this rustic retreat. Stretching over 22 feet, the porch offers sheltered involvement with outdoors and relaxing meals, and is open to the firelit living room through sliding glass doors.

**First floor—1,092 sq. ft.,
Basement—896 sq. ft.**

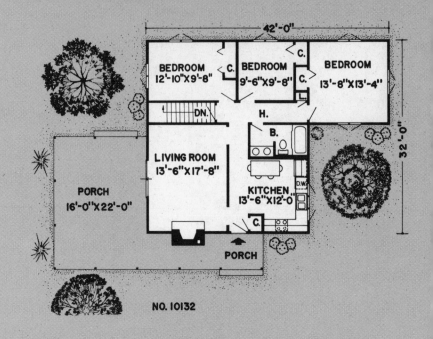

NO. 10132

Tri-level plan livable year round

No. 10222—Set to savor the scenery from four sides and three levels, this three bedroom plan does double duty as a leisure design and a home for year round living. A patio extends the basement level, while decks edge and encircle the first and second floors. For efficiency, the kitchen is large and shows built-in snack bar and deck entry. Closet space is plentiful in bedrooms and hallways, and three full compartmented baths are featured.

**First floor—1,008 sq. ft.,
Second floor—851 sq. ft.,
Basement—1,008 sq. ft.**

*For price and order information
see pages 108-109*

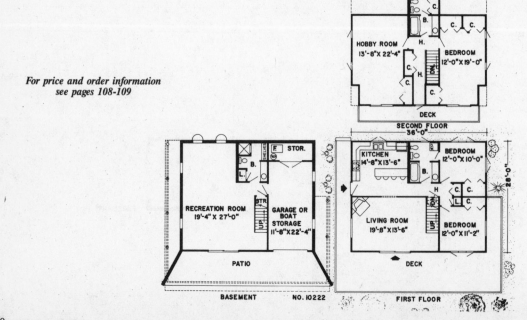

Novel design focuses on outdoor fun

No. 9890—Centering on life in the open air, this novel exterior appendages not only a balcony, open deck and terrace, but a glassed-in porch that converts to a screened-in porch for summer use. The master bedroom with private bath augments two upstairs bedrooms and a den in the basement that could generate a fourth bedroom. Also featured is a thirty foot basement recreation room flanking the terrace, plus a family room opposite the glassed-in porch.

First floor—1,280 sq. ft.,
Second floor—448 sq. ft.,
Basement—1,280 sq. ft.

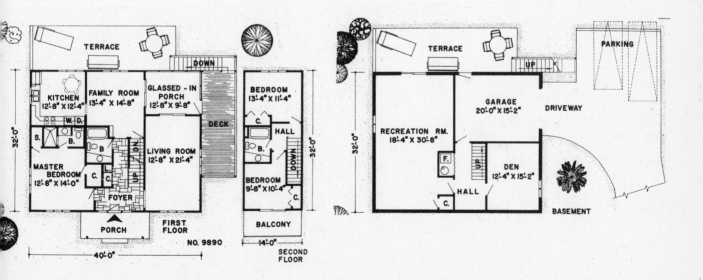

Deck enlarges, enhances cottage

No. 10306—Ideal for a beach or mountain vacation home, this one bedroom cottage supplies a large wood deck for dining, sunbathing, or relaxing with friends. The plan is strictly functional and calls for an 11-ft. bedroom adjoining a bath with shower, and a simple one wall kitchen open to the living room. Two smaller decks are also shown.

First floor—408 sq. ft.

*For price and order information
see pages 108-109*

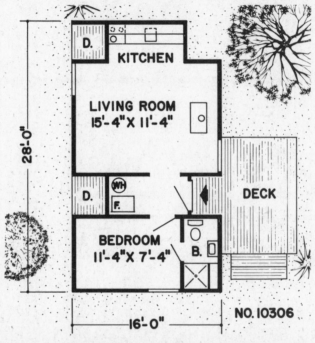
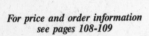

Deck encircle versatile design

No. 10324—Decks add to living, dining and bedroom areas of this two level design combining the outdoor living areas as a leisure home with the space and planning to assure livability year around. Filling the upper level are three large bedrooms, two full baths, an L shaped living-dining center and eat in kitchen. Below, family room and hobby room are steps from a third bath and laundry room.

**Upper level—1,352 sq. ft.,
Lower level—1,300 sq. ft.
(includes garage)**

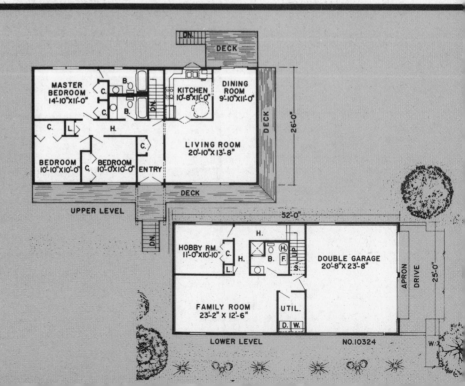

Space, privacy aims of unique bi-level

No. 10330—Two levels and four wings punctuated with decks and terraces fashion a design that stresses privacy and plenty of space. In all, the plan calls for eight totally separate areas in addition to central dining room and three baths. Living room and family room boast fireplaces and the master bedroom merits a large private bath.

Upper level—1,260 sq. ft.,
Lower level—1,260 sq. ft.

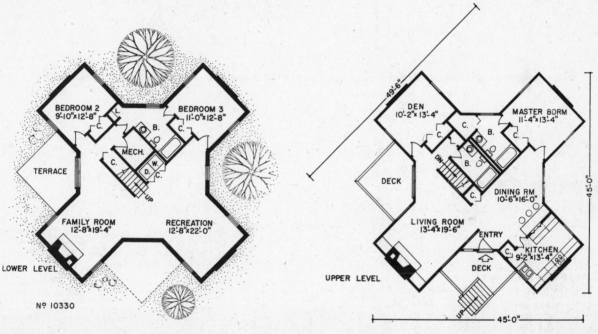

Cathedral ceilings suggest spaciousness

No. 9036—Cathedral ceilings and the lavish use of windows, especially in the living and dining room, effect the stream of light that brightens and seems to widen this contemporary ranch. The functional kitchen, open on both ends, borders the combination laundry and half bath. Toy storage, perfect for little wagons and tricycles, is provided off the carport, and a paved terrace skirts the living room on the right.

**First floor—1,108 sq. ft.; Storage—102 sq. ft.,
Terrace—200 sq. ft., Carport—542 sq. ft.**

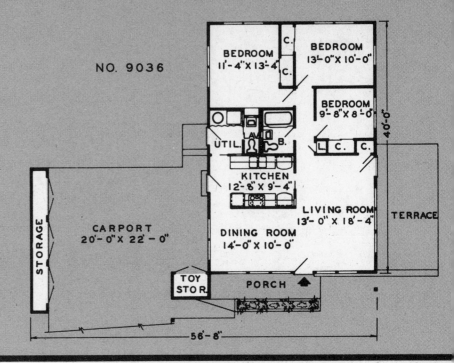

NO. 9036

Attractive roof shades terrace

No. 9564—Sheltered by the extended roof, the terrace and breezeway area will prove a tremendous setting for outdoor relaxation. You can barbecue outside in bad weather or sit on the covered terrace and enjoy a summer rain. The appealing exterior is complemented by open planning inside. The living-dining-kitchen complex spans the left wing of the home, creating an airy illusion of space. Three bedrooms include a master bedroom with private bath, and the attractively roofed garage is fringed with storage space.

First floor—1,247 sq. ft., Garage—327 sq. ft.

*For price and order information
see pages 108-109*

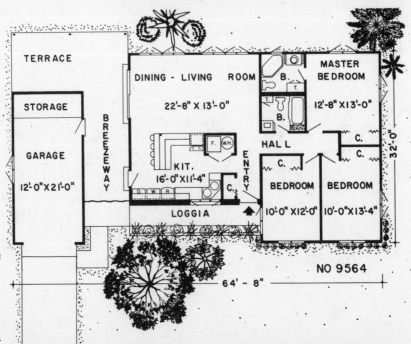

NO 9564

Functional plan specifies two full baths

No. 9038—Sharp and simple in design, this three bedroom home exhibits an economical rectangular plan that capitalizes on living space and features an extra full bath off the master bedroom. The living room and dining room together measure over 27 feet wide, and the kitchen is open on both ends, lending a sense of spaciousness to this area which is further augmented by the cathedral ceiling construction. The full basement should solve any storage problems.

**First floor—1,132 sq. ft.,
Basement—1,132 sq. ft., Carport—286 sq. ft.**

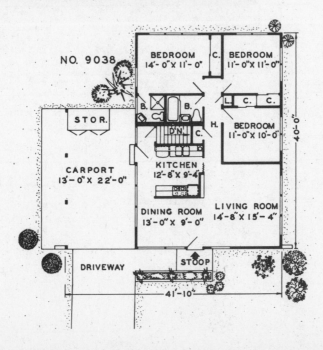

NO. 9038

BEDROOM
14'-0" X 11'-0"

C.

BEDROOM
11'-0" X 11'-0"

STOR.

B.

B.

L. C. C.

H.

BEDROOM
11'-0" X 10'-0"

CARPORT
13'-0" X 22'-0"

D'N.

C.

KITCHEN
12'-8" X 9'-4"

40'-0"

DINING ROOM
13'-0" X 9'-0"

LIVING ROOM
14'-8" X 15'-4"

DRIVEWAY

STOOP

41'-10"

Angled bedrooms distinguish exterior

No. 9904—Set at a 45° angle to the rest of the home, the two rear bedrooms contribute to a fascinating and eye-catching deviation from the traditional rectangular ranch style. In all, four bedrooms, nine closets and two baths comprise the right wing of the home. Fireplaces favor both the living room and family room, and the attractive bow window in the dining room looks out on the immense terrace. Another half bath lies off the family room and includes space for the washer and dryer.

**First floor—2,565 sq. ft.,
Basement—2,565 sq. ft., Garage—528 sq. ft.**

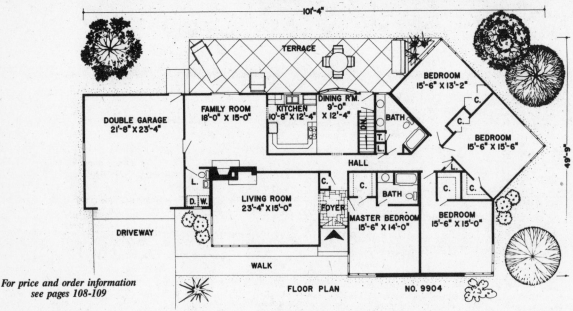

*For price and order information
see pages 108-109*

FLOOR PLAN   NO. 9904

Small design exudes character

No. 10072—Ornamented garage door, shutters, and cupola add interest o the exterior of this compact plan, while the interior expresses a personality of its own, sparked by angled garage and firelit living room. To the left of the foyer, functional corridor kitchen saves steps and serves the dining area, and the main living room enjoys wood-burning fireplace and access to the patio. Plentiful storage space fringes the double garage, and three well-closeted bedrooms share a large bath.

**First floor—1,056 sq. ft.,
Basement—1,056 sq. ft.,
Garage—633 sq. ft.**

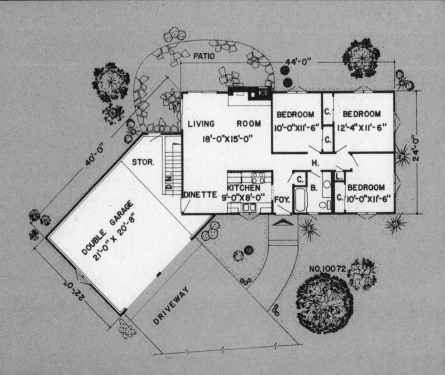

Studio crowns enchanting design

No. 10068—Reached by a spiral staircase, the balconied studio level including bedroom and full bath tops this unique design. Inside the split foyer arrangement allows separation of noise areas from quiet zones and places three sizable bedrooms, formal living and dining rooms and kitchen on the upper level.

Main level—1,560 sq. ft.,
Lower level—1,560 sq. ft.,
Studio level—480

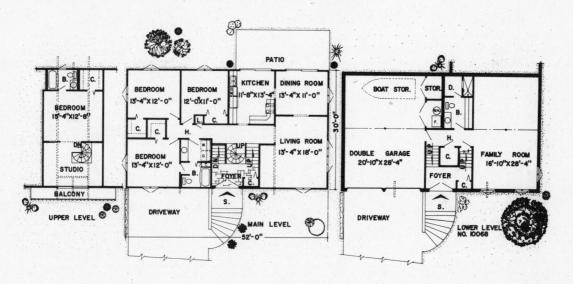

Elevated sun deck savors scenery

No. 10058—High enough to capture a sweeping view, the wooden deck joins the sliding glass doors in permeating the living room with light and scenery. The open living, dining and kitchen area boasts a fireplace and breakfast bar. Besides the bedroom and bath on the upper level, two more bedrooms, a bath, and a laundry room comprise the lower level. Housed in the fireplace foundation in the carport is a built-in barbecue, serviceable in any weather.

Lower level—636 sq. ft.,
Upper level—768 sq. ft., Carport—264 sq. ft.

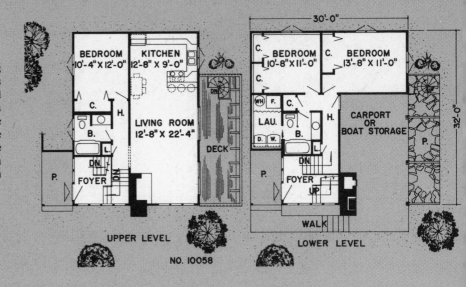

NO. 10058

Outdoor dining slated for deck

No. 10182—Dining in the fresh air is encouraged by the 12-foot outdoor dining room, equipped with built-in barbecue grill, and part of the massive deck area in this rustic leisure plan. Two pairs of sliding glass doors separate the deck from the substantial living room, highlighted by wood-burning fireplace. Ample storage space serves the first floor bedroom, which is bordered by full bath and laundry. Upstairs, another large bedroom and an all purpose room promise sleeping quarters for family and visiting friends.

First floor—968 sq. ft.,
Second floor—640 sq. ft.,
Carport—264 sq. ft.

*For price and order information
see pages 108-109*

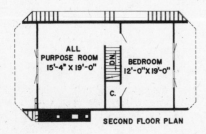

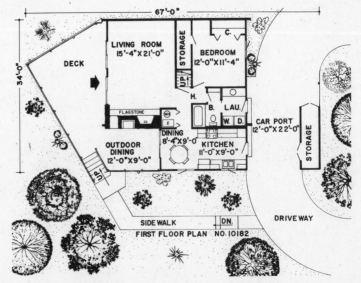

Home reflects Alpine image

No. 10120—Reminiscent of a Swiss chalet, this home features an open deck and lower-level double garage. Complete with basement, the home combines architectural beauty with convenience. The outside staircase opens into a foyer which channels traffic either into the living room or kitchen or hallway.

First floor—1,792 sq. ft.,
Second floor—523 sq. ft.,
Basement garage—859 sq. ft.,
Basement—709 sq. ft.

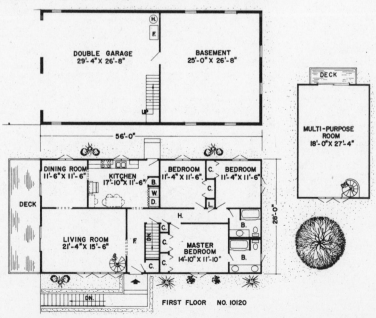

FIRST FLOOR   NO. 10120

Penthouse basks in luxury of own sun deck

No. 9648—Singular and striking, this upper story penthouse is streamed with light through expanses of glass and enriched with a sprawling sun deck. The level beneath is not to be outdone, and presents a three bedroom layout awarding the master bedroom a private bath with luxurious corner bathtub. The living room with fireplace is formal and extensive and is balanced by a casual family room connecting to the concrete terrace. A double garage includes an entrance to the kitchen.

First floor—1,575 sq. ft.,
Basement—1,575 sq. ft.,
Second floor—342 sq. ft., Garage—433 sq. ft.

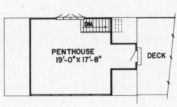

For price and order information see pages 108-109

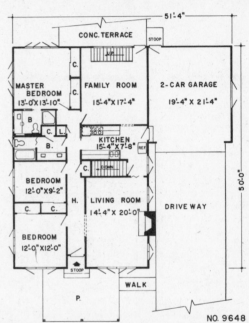

Angular design stresses family area

No. 10236—Accessible from all three sides, the triangular family area, incorporating kitchen and dining space, forms the focal point of this unusual vacation home. The design, eye-catching and efficient, offers pantry, plenty of storage space, and built-in shelves. Directly off the family area is the master bedroom, large and double-closeted, and bordering full bath. Two upstairs bedrooms open to outdoor deck and indoor balcony.

First floor—805 sq. ft.,
Second floor—445 sq. ft.,
Carport—192 sq. ft.,
Outside storage—64 sq. ft.

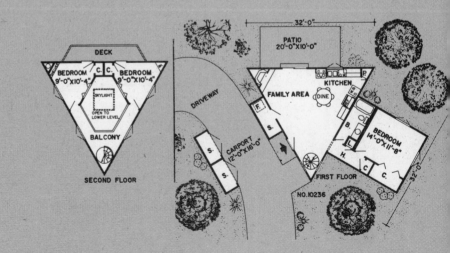

Ceilings open design to sunlight

No. 9750—Sunlight streams into the family room of this sharp contemporary via lofty cathedral ceilings that assure light, privacy and an aura of spaciousness. Bordering the family room is a kitchen with access to the patio and to the laundry/utility room behind the garage. Three bedrooms are outlined, including a master bedroom with half-bath, and closet space is adequate. A well-lighted living room provides a natural setting for entertaining.

First floor—1,156 sq. ft.,
Garage—484 sq. ft.

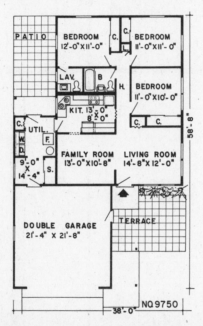

Exciting living . . . American style!

No. 10106—Beamed ceiling and fireplace highlight this contemporary plan's living room on the upper floor shared only by library and bedrooms. Below, another fireplace warms the family room, outlined for activity and merging with dining area and corridor kitchen. Laundry center within the kitchen area is a convenient extra. Baths on both floors are calculated for utmost efficiency and incorporate double entrances, dressing rooms, and compartmented privacy.

Lower level—1,354 sq. ft.,
Upper level—1,354 sq. ft.,
Garage—560 sq. ft.

Skylit plan set for beach living

No. 9392—Central skylight, expansive sun deck, and wooden piers suit this hexagonal contemporary to carefree beach living. On the entry level, a closet, full bath with shower, and laundry center simplify cleaning up after a day in the sand. Up the circular staircase, two bedrooms allow for ample sleeping, storage and closet space, and are steps from a second full bath. Sliding glass doors on two sides of the living room and in the kitchen-family room provide light and access to the roomy sun deck that encircles the plan.

Upper level—1,038 sq. ft.,
Lower level—261 sq. ft.

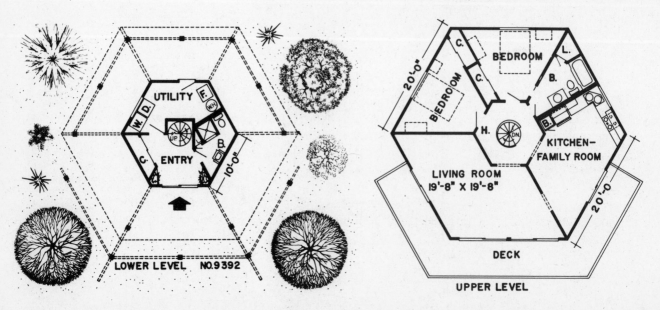

U-shaped plan encloses outdoors

No. 10094—Rough plywood siding and shake shingle mansard roof blend this design with the countryside; its U-shaped styling brings the outdoors inside. The plan includes three sizable bedrooms, full bath, kitchen and living room, all encircling 245 square feet of private patio. The full bath is central, and a wood-burning fireplace cheers the home's entire right wing. Master bedroom and living room share the patio through sliding glass doors.

House—1,338 sq. ft., Patio—245 sq. ft.

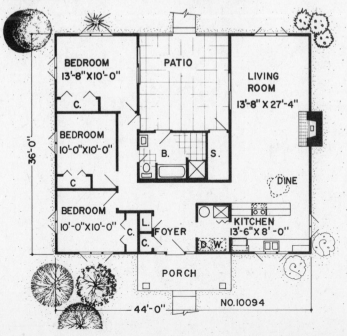

*For price and order information
see pages 108-109*

BEDROOM
13'-8"X10'-0"
C.

BEDROOM
10'-0"X10'-0"
C

BEDROOM
10'-0"X10'-0"
C.

PATIO

LIVING ROOM
13'-8" X 27'-4"

B.   S.

DINE

L. C.   FOYER
C.

KITCHEN
13'-6"X 8'-0"

D.   W.

PORCH

36'-0"

44'-0"

NO. 10094

Glass captures views & sun in A-frame

No. 21100—Abundant glass floods this plan with light and offers images of the surrounding scenery from three sides as well as serving as a solar energy feature. Large exterior exposed beams crisscross the glass giving a massive, rugged appearance. The center of family activity begins in the family room and proceeds to the deck through sliding glass doors. Wooden seats rail the deck which flows into a dining patio on the left side. Your family may relax over meals here or in the dining/kitchen area just inside glass doors. Two bedrooms, a full bath and laundry facilities complete the first level. An open wooden stairway channels you to the second level which opens into a large fireplaced sitting room and balcony overlooking the family room. Full length windows frame both ends of the sitting room. A very private master bedroom with full bath occupies the entire back of the upper level. Plans for optional basement, crawl space or concrete slab are provided.

First Floor—1,126 sq. ft.
Second Floor—603 sq. ft.

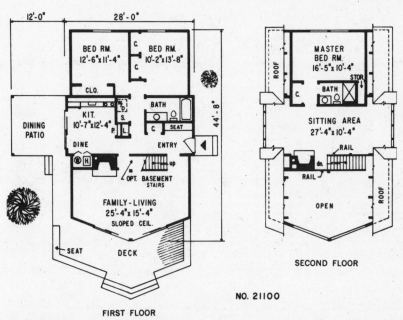

For price and order information
see pages 108-109

Old American saltbox design

No. 21104—A sloping living room ceiling creates a sense of spaciousness to the modest square footage. You can relax in front of the centrally located fireplace in cool weather or move through triple sliding glass doors to the roomy deck when the weather is warmer. Behind the living room lies a bedroom, full bath and kitchen/dining area which has a window seat. Laundry facilities are conveniently placed off the kitchen. On the left of the living room a quiet corner has been tucked under the stairs leading to the second floor. The second level affords two equal sized bedrooms (one with its own private deck), joined by a full bath. A balcony skirts the entire level and overlooks the living room below.

First Floor—840 sq. ft.
Second Floor—440 sq. ft.

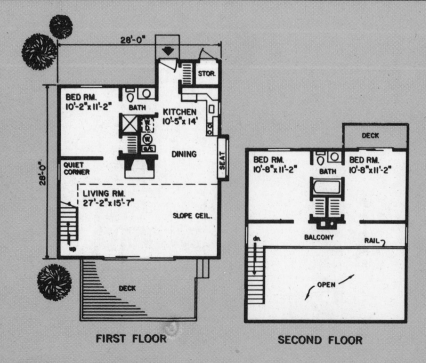

44

Spiral stairs is focal point

No. 21102—Wood shakes and wood siding create a rusticness that thrusts this salt box design deep into your favorite vacation spot. A centrally located open spiral staircase circles upward to form a focal point of the floor plan. A living/dining room with fireplace and log storage lies to the front and overlooks the seat rimmed deck from sliding glass doors and windows. The deck, dining area, and eating bar next to the kitchen offer a wide range in both formal and informal dining. A versatile ski storage/ski hall/laundry area, full bath and bedroom finish out the first level. The water heater and furnace are tucked between the living room fireplace and bedroom. Two bedrooms and a full bath occupy the second level.

First Floor—780 sq. ft.
Second Floor—437 sq. ft.

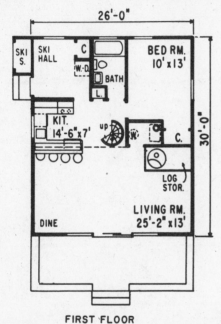

FIRST FLOOR

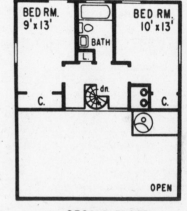

SECOND FLOOR

NO. 21102

An entry atrium

No. 26811—This is a house for people who enjoy being surrounded by nature. The entrance foyer sets the mood. The 12-by-18 foot, two-story, skylit space can function as a garden room. From the atrium, a double door opens into an L-shaped hallway leading to a corridor kitchen, a big square dining room, and a sunken living room. Two steps up from the living room is the owners luxurious suite. Exercise and relaxation are well provided for in all these rooms. The dining room and living room both open to a redwood deck where a hot tub can be installed; the master suite has a separate room for exercising. Two additional bedrooms and a full bathroom are located on the lower level, close to a recreation room, patio and laundry.

Main level—1,713 sq. ft.,
Lower level—1,425 sq. ft.,
Garage—676 sq. ft.

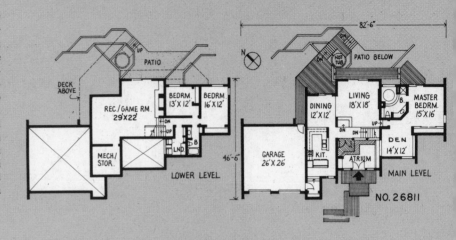

Entry a year-round solar greenhouse

No. 26720—Providing a dramatic entry and yielding passive solar benefits, a 16-foot-long greenhouse sparkles at the center of this rough-sawn fir-plywood exterior. Glass doors open to feed solar gain to the house, allowing heat to circulate through the open-plan first-floor. Convection carries heat to second-floor bedrooms, and circulation might be improved by building the master bedroom as an open balcony. Where the living room ceiling rises two stories, a heat-circulating fireplace creates a focus for seating. Nearby, a low ceiling sets off a dining area, and an open kitchen doubles as a social area that never isolates the cook from guests. An open stair ascends in a half turn to the second floor, where the master suite and two smaller bedrooms are divided by a sound-muffling core of bathrooms and closets.

First floor—845 sq. ft.,
Second floor—1,105 sq. ft.,
Greenhouse—160 sq. ft., Basement—855 sq. ft.,
Garage—516 sq. ft.
Richard Eschliman—Architect

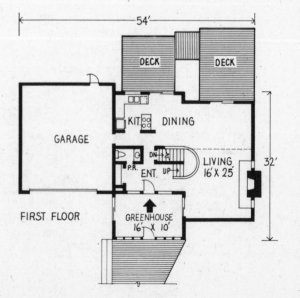

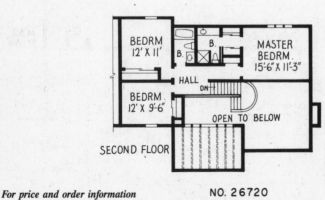

*For price and order information
see pages 108-109*

NO. 26720

Redwood-sheathed, passive solar

No. 26900—This two-story, south-facing wood and glass house was designed for passive solar heating, and it also can, because of the slope of its roof and its plumbing system, be fitted with active roof collectors for the preheating of domestic hot water as well. A two-story, glass-walled entry captures the sun's heat, which it will radiate to adjacent rooms. The solar entry, which might become a greenhouse, leads into the Great Room or den. The Great Room, over 24 feet long, separates the kitchen on the left from a master suite on the right. The kitchen is designed to act as the informal hub with its U-plan work area open to a breakfast area. On the upper level, there are two large bedrooms, a compartmented bathroom, and a loft over the master bedroom.

First level—1,504 sq. ft.,
Second level—712 sq. ft., Garage—400 sq. ft.,
Basement—400 sq. ft.

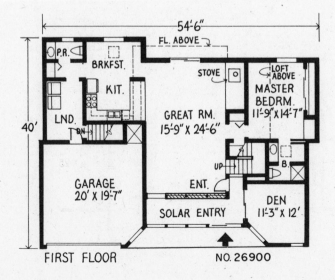

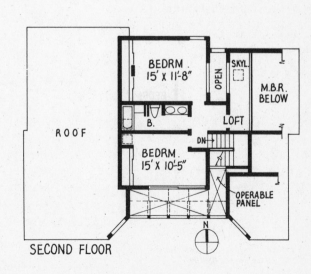

47

Master suite on a private level

No. 26810—Its four floors staggered at half-level intervals, this house is both architecturally fascinating and effectively planned. Entering on the third level, one sees dining and sunken living rooms ahead on a space-expanded diagonal. The corridor kitchen extends into a traffic-free space open to living areas on one side, and a deck makes the outdoors a natural part of all social areas. One half-level higher, the master bedroom connects to a study and deck plus a luxurious compartmented bathroom. On the second level, two smaller bedrooms have a landing with a bath and convenient laundry. The lowest level of the house is a recreation basement. Framing of this house uses large studs and rafters spaced at wide intervals to cut construction time, reduce the need for lumber and open deeper gaps for thicker insulating batts.

**Upper level—1,423 sq. ft.,
Lower level—1,420 sq. ft., Garage—478 sq. ft.**

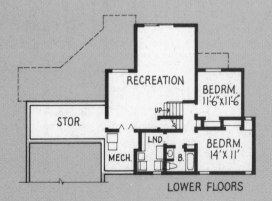

48

Exceptional retreat

No. 26602—Just the thing for a hunting
lodge, summer/winter cabin or family
retreat. This economical energy efficient,
open living design features unique solar
advantages. Entire south wall is made of
glass which gains direct heat from the
sun while two full length thermal walls
release heat slowly at night. For cheer-
fulness and additional heat a "free stan-
ding" fireplace is located in the center.
Hinged walls on the front and back help
control the solar effectiveness of this
charming little retreat. A sleeping loft
located at one end is made accessible by
way of a ladder.

**Main level—440 sq. ft.,
Sleeping loft—112 sq. ft.**

OPEN AREA

OPEN | **SLEEPING LOFT** | OPEN

34'-0"

PATIO

34'-0"

B.

S.

FIREPLACE

KITCHEN

STORAGE

PATIO

FIRST FLOOR   NO. 26602

*For price and order information
see pages 108-109*

Split foyer governs unique plan

No. 10254—Placing bedrooms on the lower level and living rooms above, this exceptional design uses its unique exterior to capture light for living areas and create a plan that is highly workable. Bedrooms are large and fill the lower level; living areas are unrestricted and accented with balcony and wood-burning fireplace. A snack bar divides kitchen and dining room.

Upper level—784 sq. ft.,
Lower level—784 sq. ft.

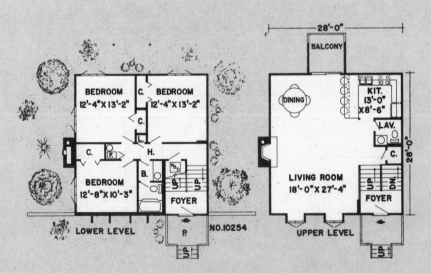

LOWER LEVEL   NO. 10254   UPPER LEVEL

BEDROOM 12'-4" X 13'-2"

BEDROOM 12'-4" X 13'-2"

BEDROOM 12'-8" X 10'-3"

FOYER

28'-0"

BALCONY

DINING

KIT. 13'-0" X 8'-6"

LAV.

LIVING ROOM 18'-0" X 27'-4"

FOYER

28'-0"

Small but versatile passive design

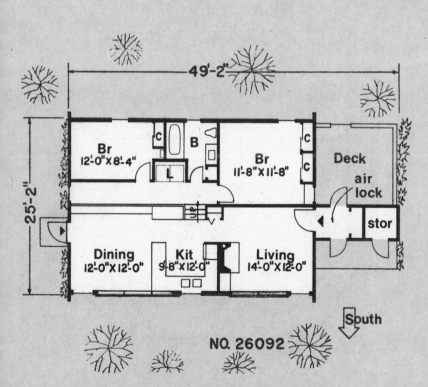

49'-2"

25'-2"

Br
12'-0"x8'-4"

B

Br
11'-8"x11'-8"

Deck
air
lock

stor

Dining
12'-0"x12'-0"

Kit
9'-8"x12'-0"

Living
14'-0"x12'-0"

South

NO. 26092

No. 26092—This house is well suited as a vacation home or for a small family. The One Design Waterwall passive solar system is employed in the southern walls. R-23 insulation is used in exterior walls, R-19 in the floors and R-30 in the ceilings. An air lock entry and coniferous trees on the north side for a winter windbreak further add to energy efficiency. Clerestory windows allow the sun's warmth to enter. The entryway with adjoining storage space directs you into the fireplaced living room or out onto the deck. A kitchen and dining room are also on this level. Two bedrooms and a full bath lie up several steps.

Living area—925 sq. ft.
Entry & Storage—80 sq. ft.

*For price and order information
see pages 108-109*

51

Passive contemporary design features sunken living room

No. 26112—Wood adds its warmth to the contemporary features of this passive design. Generous use of southern glass doors and windows, an air lock entry, skylights and a living room fireplace reduce energy needs. R-26 insulation is shown for floors and sloping ceilings. Decking rims the front of the home and gives access through sliding glass doors to a bedroom/den area and living room. The dining room lies up several steps from the living room and is separated from it by a half wall. The dining room flows into the kitchen through an eating bar. A second floor landing balcony overlooks the living room. Two bedrooms, one with its own private deck, and a full bath finish the second level.

First floor—911 sq. ft.
Second floor—560 sq. ft.

Leisure-oriented plan features sauna

No. 10258—Sauna, powder room, and shower room, all accessible from the covered deck, set the pace for this fun-oriented four bedroom home. Ideal for a setting near lake or ocean, the design shows a main level master bedroom, with three additional bedrooms accessible via a circular stairway. On the lower level, the recreation room sports a built-in bar, and a multi-purpose room and closeted half bath are featured.

First floor—1,403 sq. ft.,
Second floor—1,035 sq. ft.,
Lower level—776 sq. ft., Garage—624 sq. ft.

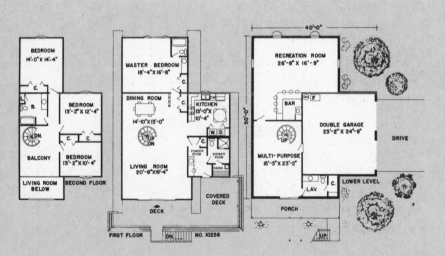

*For price and order information
see pages 108-109*

53

Home adapts to year round, leisure use

No. 10244—With storage space for a recreational vehicle, careful attention to outdoor living areas, and a handy kitchenette off the living room, this three bedroom design adapts readily to any situation. A large family room furnishes the first level, while a living room with deck is found on the upper level. Two and one half baths plus a laundry center are provided.

**First floor—1,050 sq. ft.,
Second floor—798 sq. ft.,
Garage—606 sq. ft.**

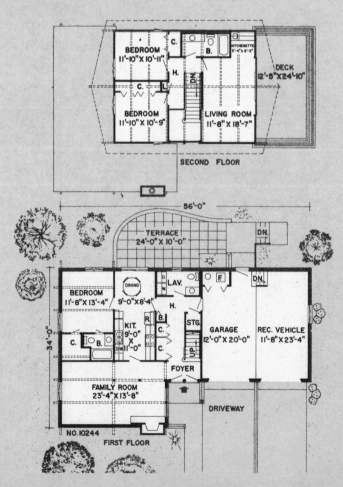

SECOND FLOOR

BEDROOM
11'-10" X 10'-11"

BEDROOM
11'-10" X 10'-9"

LIVING ROOM
11'-8" X 18'-7"

KITCHENETTE
5'-4" X 5'-3"

DECK
12'-5" X 24'-10"

56'-0"

TERRACE
24'-0" X 10'-0"

BEDROOM
11'-8" X 13'-4"

DINING
9'-0" X 8'-4"

LAV.

KIT.
9'-0" X 11'-0"

STG.

GARAGE
12'-0" X 20'-0"

REC. VEHICLE
11'-8" X 23'-4"

FOYER

FAMILY ROOM
23'-4" X 13'-8"

34'-0"

DRIVEWAY

NO. 10244

FIRST FLOOR

Double-closeted bedrooms highlight plan

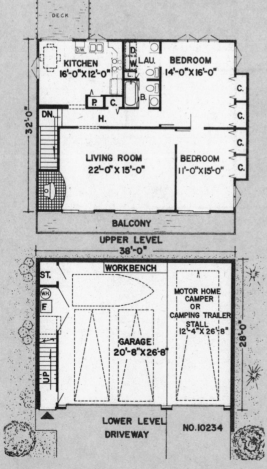

KITCHEN
16'-0" X 12'-0"

D.W.

LAU.
D W

BEDROOM
14'-0" X 16'-0"

B.

P. C.

DN.

H.

LIVING ROOM
22'-0" X 15'-0"

BEDROOM
11'-0" X 15'-0"

C.

C.

C.

C.

32'-0"

DECK

BALCONY

UPPER LEVEL

38'-0"

WORKBENCH

ST.

W.H.

F

GARAGE
20'-8" X 26'-8"

MOTOR HOME
CAMPER
OR
CAMPING TRAILER
STALL
12'-4" X 26'-8"

UP

28'-0"

LOWER LEVEL
DRIVEWAY

NO. 10234

No. 10234—Generously proportioned bedrooms, with double closets and bordering laundry and full bath, promise livability in this trim leisure plan. For entertaining, the 22-ft. living room expands via two sets of sliding glass doors to the large balcony. A spacious kitchen offers dining space and pantry and opens to the wooden deck for easily-prepared outdoor meals. Lower level shows garage, workbench, and storage for camper or trailer.

Upper level—1,254 sq. ft.,
Lower level—1,064 sq. ft.

*For price and order information
see pages 108-109*

Multi-level contemporary

No. 26111—The features of this multi-level contemporary home lend character to both the exterior and interior. A wooden deck skirts most of three sides. Great variety in the size and shape of doors and windows is apparent. Inside the living room forms a unique living center. It can be reached from sliding glass doors from the deck or down several steps from the main living level inside. It is overlooked by a low balcony from the entryway and dining room on the lower level and from the second floor landing. Large windows on both the right and left keep it well lit. A fireplace here is optional Ceilings slope upward two stories. A partial basement is located below the design.

First floor—769 sq. ft.
Second floor—572 sq. ft.

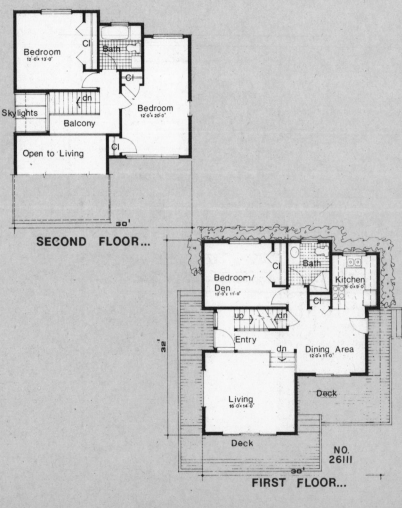

SECOND FLOOR...

FIRST FLOOR...

Plan spotlights two story living room

No. 10300—Decks, balconies, and an impressive two story living room with fireplace make this design a vacation home for all seasons. Get away from it all or entertain in style in the living area or well-equipped kitchen-dining area. Indoor balcony overlooks the living room, while outdoor decks or balconies serve each of the bedrooms. Laundry space is featured.

First level—507 sq. ft.,
Second level—1,095 sq. ft.,
Third level—949 sq. ft.

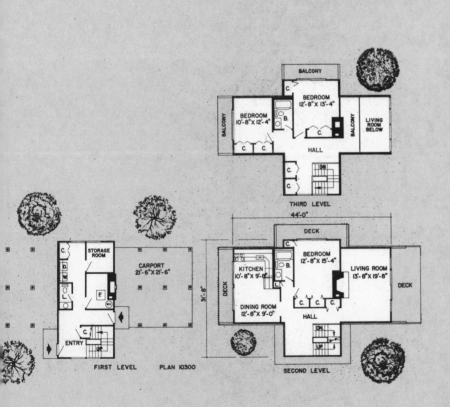

For price and order information see pages 108-109

Living room focus of spacious home

No. 10328—Equipped with fireplace and sliding glass doors to the bordering deck, the two story living room creates a sizable and airy center for family activity. A well planned traffic pattern connects dining area, kitchen, laundry niche and bath. Closets are plentiful and a total of three 15 foot bedrooms are shown.

First floor—1,024 sq. ft.,
Second floor—576 sq. ft.,
Basement—1,024 sq. ft.

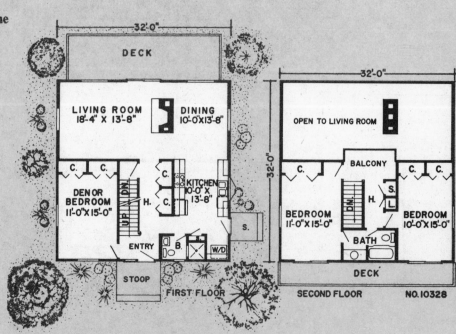

Passive solar and contemporary features

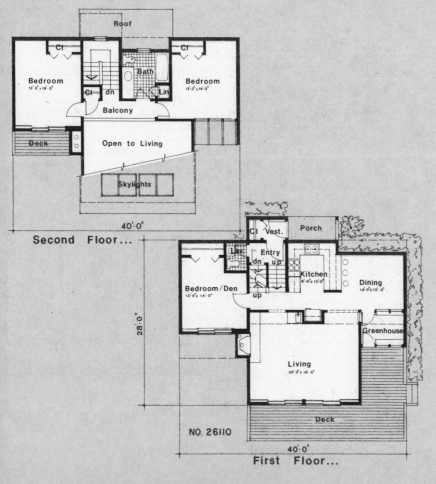

Roof

Cl | Cl

Bedroom
12'0" x 16'0"

Bath

Bedroom
12'0" x 14'0"

Cl | dn | Lin

Balcony

Deck

Open to Living

Skylights

40'-0"

Second Floor...

Cl | Vest. | Porch

Entry

dn | up

Kitchen
9'0" x 12'0"

Bedroom/Den
12'0" x 14'0"

up

Dining
12'0" x 10'0"

28'-0"

Greenhouse

Living
20'0" x 16'0"

NO. 26110

Deck

40'-0"

First Floor...

No. 26110—Numerous southern glass doors and windows, skylights and a greenhouse clue the exterior viewer of the passive solarness of this contemporary design. For minimum heat loss, 2x6 studs for R-19 insulation are shown in exterior walls and R-33 insulation is shown in all sloping ceilings. The living room employs a concrete slab floor for solar gain. Basement space is located under the kitchen, dining room, lower bedroom and den. A northern entrance through a verstibule and French doors channels you upward to the first floor living area. A unique feature on this level is the skylit living room ceiling which slants two stories. Second story rooms are lit by clerestory windows. Two balconies are on this level, and exterior one off the bedroom and an interior one overlooking the living room.

First floor—902 sq. ft.
Second floor-567 sq. ft.

*For price and order information
see pages 108-109*

Natural elements mingle in modern plan

No. 10184—Rustic beams, exposed
stone, and a cedar deck combine to
allow the mingling of indoors and outdoors in this unusual design. Wood-
burning fireplaces grace the living room
and covered patio, and clerestory windows over the living room promise
abundant light and atmosphere.
Separated from living areas by an exposed stone wall, the sleeping wing houses
two bedrooms and two baths, with
another bedroom a possibility on the
lower level. The eat-in kitchen opens to
the cedar deck via sliding glass doors.

Upper level—1,500 sq. ft.,
Lower level—619 sq. ft.,
Carport and storage—870 sq. ft.

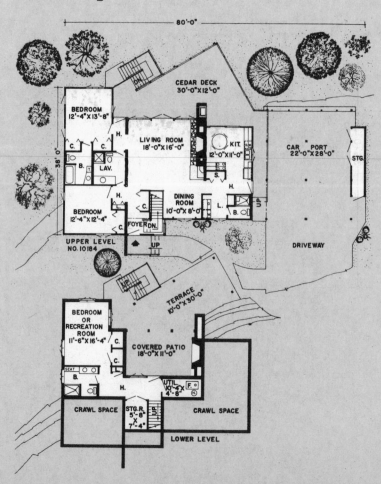

Deck encircles compact home

No. 19709—Suitable for a vacation or small family residence, this modest two bedroom home provides a spacious living area and a comfortable deck for outdoor relaxation. Free flow of space in dining area, living area, and kitchen make the areas seem larger, and two sets of sliding glass doors provide light and access to the deck. The kitchen pantry is a bonus.

First floor—792 sq. ft.,
Deck—588 sq. ft.

*For price and order information
see pages 108-109*

FLOOR PLAN

22'-0"

36'-0"

BDRM 11½ x 10½

BDRM 11½ x 10½

DINING 13 x 8

LIVING 13 x 13

DECK 11 x 27

Leisure home sports party, guest space

No. 10226—With a second level that features lounge, kitchenette, and deck, as well as bunk space, this one bedroom home creates a retreat for family and guests. Weekend living focuses on the main level, where expansive L-shaped living room flows easily into the corridor kitchen. Living room and bedroom open to the 30-ft. deck, and compartmented bath efficiently serves living and sleeping areas. Foyer and basement are bonuses.

First floor—884 sq. ft.,
Second floor—546 sq. ft.,
Basement—452 sq. ft.,
Carport & Storage—400 sq. ft.

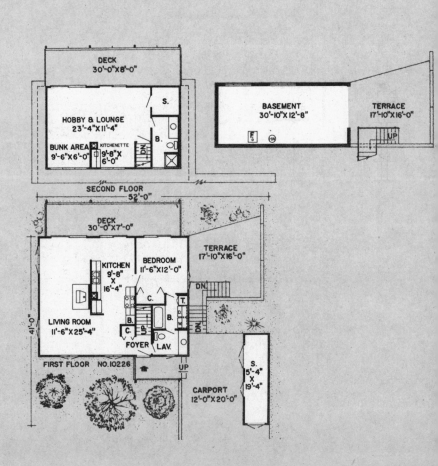

Vacation home covers all angles

No. 10296—Expanses of glass jut out in all four directions to assure a beautiful view from this two bedroom leisure home. Entry is via the patio/carport level, where the second bedroom sports two closets and adjoins a full bath. Notable are the sliding glass doors from each bedroom, the balconied living room with fireplace, and the laundry center on the lower level.

Upper level—1,024 sq. ft.,
Lower level—512 sq. ft.

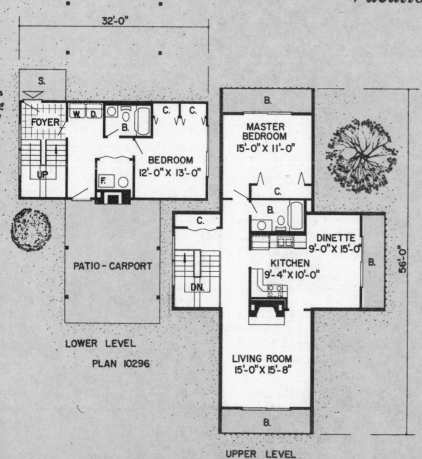

32'-0"

S.
FOYER
W. D.
B.
C. C.
UP
BEDROOM
12'-0" X 13'-0"
F.

PATIO - CARPORT

LOWER LEVEL

PLAN 10296

B.
MASTER BEDROOM
15'-0" X 11'-0"
C.
B.
C.
DINETTE
9'-0" X 15'-0"
KITCHEN
9'-4" X 10'-0"
B.
DN.
LIVING ROOM
15'-0" X 15'-8"
B.

56'-0"

UPPER LEVEL

*For price and order information
see pages 108-109*

A heat-gathering garden room

No. 26820—One of the smallest rooms in this house, the garden room, might turn out to be the most important. Oriented south or southwest, its skylight and insulating glass wall gather sunlight that will not only make plants thrive but will also help heat the house. Sliding glass doors can be opened to draw this heat into the living areas to supplement the conventional mechanical system. The design guarantees many pleasures for outdoor enthusiasts: two decks, skylighting, year-around green views of the garden room, even a potting room by the back door. The core of the plan is a "keeping room" combining kitchen, breakfast area and family area by a fireplace. Each living area has three exposures for exhilarating light and summer comfort.

First level—1,618 sq. ft.,
Second level—907 sq. ft.,
Basement—1,621 sq. ft.,
Garage—552 sq. ft.,
Garden room—98 sq. ft.

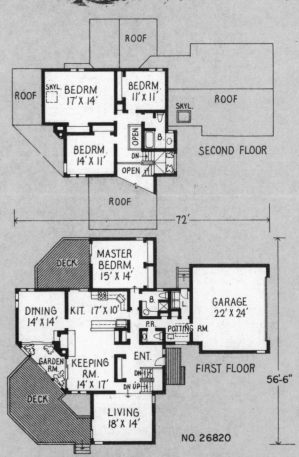

Excellent zoning marks A-frame

No. 10228—Besides separating bedrooms from living areas by placing them on different floors, this rustic A-frame plan also uses the entry and half bath to divide the family-kitchen and adjoining patio from the guest-oriented living room. Dining space overlooks the patio through sliding glass doors. Upstairs, two spacious bedrooms are indulged with private decks and are served by full bath and linen closet.

First floor—768 sq. ft.,
Second floor—521 sq. ft.

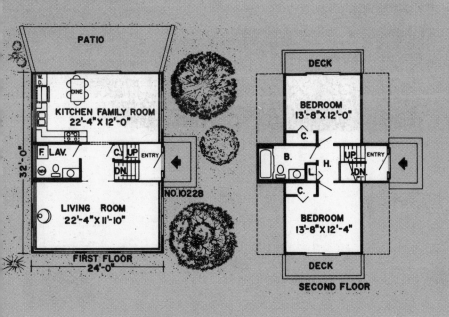

*For price and order information
see pages 108-109*

Space for group activities

No. 9708—The sundeck is large enough to be useful and is partially under roof to provide both sunny and shady areas. Access to the sundeck is through sliding glass doors in both the living room and family room. An excellent floor plan provides three bedrooms and two full baths. The baths are centrally located for maximum efficiency. The basement floor plan can be arranged to suit individual preference and family requirements.

First floor—1,636 sq. ft.,
Basement—1,636 sq. ft.,
Garage—523 sq. ft.

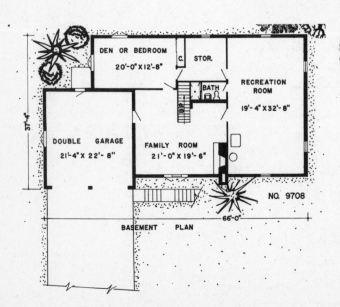

DEN OR BEDROOM
20'-0" X 12'-8"

STOR.

C.

BATH

RECREATION ROOM
19'-4" X 32'-8"

37'-4"

DOUBLE GARAGE
21'-4" X 22'-8"

FAMILY ROOM
21'-0" X 19'-6"

NO. 9708

66'-0"

BASEMENT PLAN

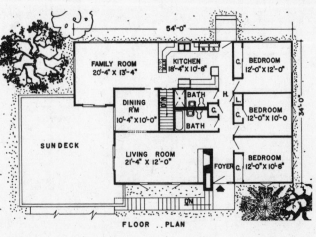

54'-0"

FAMILY ROOM
20'-4" X 13'-4"

KITCHEN
18'-4" X 10'-8"

BEDROOM
12'-0" X 12'-0"

DINING R'M
10'-4" X 10'-0"

BATH

H.

C.

BATH

BEDROOM
12'-0" X 10'-0"

34'-0"

SUN DECK

LIVING ROOM
21'-4" X 12'-0"

FOYER

C.

BEDROOM
12'-0" X 10'-8"

DN

FLOOR PLAN

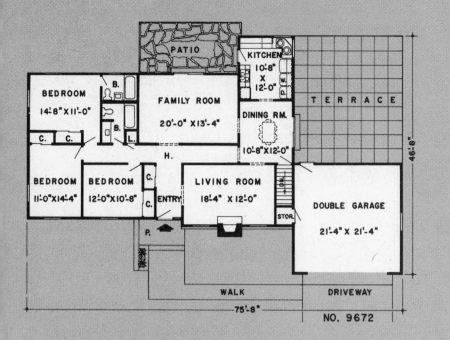

PATIO

BEDROOM
14'-8" X 11'-0"

FAMILY ROOM
20'-0" X 13'-4"

KITCHEN
10'-8" X 12'-0"

W.
D.

TERRACE

DINING RM.
10'-8" X 12'-0"

B.

BEDROOM
11'-0" X 14'-4"

BEDROOM
12'-0" X 10'-8"

C.

ENTRY

LIVING ROOM
18'-4" X 12'-0"

DOUBLE GARAGE
21'-4" X 21'-4"

STOR.

P.

46'-8"

WALK

DRIVEWAY

75'-8"

NO. 9672

Outstanding floor plan

No. 9672—The living room and family room can be alternated if desired. This will please the housewife who likes to rearrange the furniture because in this house she can change it from one room to the other. An adequate size dining room is provided to prevent the family room from doubling as a dining room. The kitchen is well equipped and has space for both washer and dryer. The concrete terrace and flagstone patio provide plenty of space for outdoor living.

**First floor—1,671 sq. ft.,
Basement—1,671 sq. ft., Garage—477 sq. ft.**

Omitting carport suits to narrow lot

No. 9064—Neat and contemporary in design, this home is only 24 feet wide without the carport and would adapt gracefully to a nrrow lot. The technique of combining horizontal and vertical siding, and the addition of the low-pitched roof extending over the carport, make it appear larger than its 864 square feet. Two adequate bedrooms sharing the full bath make it suitable for a retired couple, who would then enjoy a guest room or den. The living room with coat closet, kitchen, dining room and full basement round out the floor plan.

**First floor—864 sq. ft, Basement—864 sq. ft.,
Carport—350 sq. ft.**

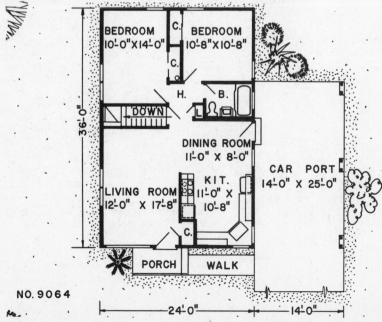

BEDROOM
10'-0" X 14'-0"

BEDROOM
10'-8" X 10'-8"

C.

DOWN

H.

B.

DINING ROOM
11'-0" X 8'-0"

CAR PORT
14'-0" X 25'-0"

LIVING ROOM
12'-0" X 17'-8"

KIT.
11'-0" X 10'-8"

C.

36'-0"

PORCH

WALK

NO. 9064

24'-0"

14'-0"

*For price and order information
see pages 108-109*

It only looks expensive

No. 250—This contemporary ranch house looks like a large expensive house, but it contains only 1,179 square feet. The living room is quite large and note the large wall areas for furniture placement. The bathroom is centrally located for easy access from all rooms. The terrace area outside the family room will make an excellent outdoor living and play area. The kitchen is a housewife's dream with washer, dryer, built-in range and oven and lots of cabinets.

First floor—1,179 sq. ft.

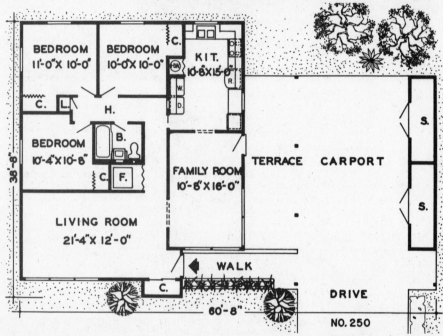

Cedar shakes both rustic, practical

No. 9858—Red cedar shake shingles that will weather naturally make this design appealing and practical. In the foreground is a large wooden deck which allows entry to the living room. A corner wood-burning fireplace adds a rustic warmth indoors. Two bedrooms, with adequate closet space, a full bath, and kitchen with eating space round out the floor plan. Outside, the attractive shed roof fashions a carport and boatport, with an additional closed storage area along the wall.

First floor—832 sq. ft.,
Storage, carport, boatport—448 sq. ft.

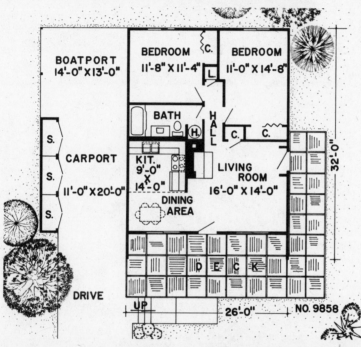

Not an inch of wasted space

No. 184—This is a small, compact plan, good for a narrow lot. The vertical siding gives the exterior an attractive appearance, while the rectangular shape and built-up roof add to the economy of this plan. There are three bedrooms, one and one-half baths, a good sized living room with fireplace and eating space in the kitchen. The plan is without a basement.

First floor—1,244 sq. ft., Garage—290 sq. ft.

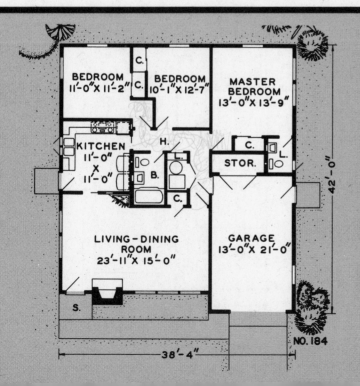

*For price and order information
see pages 108-109*

Bedrooms share spacious sun deck

No. 244—Complete relaxation and sunbathing in privacy are goals of the 30 foot sun deck stretching from three bedrooms of this cypress and brick split level. Bedrooms are favored with copious closet space, and a sizable lower level den offers additional sleeping possibilities. Tremendous storage space separates carport and terrace and is accessible to both, while the recreation room borders a half bath and opens to terrace via sliding glass doors. Open kitchen and dining room are featured.

**Living Levels—1,213 sq. ft.,
Recreation level—641 sq. ft.,
Carport and storage—373 sq. ft.**

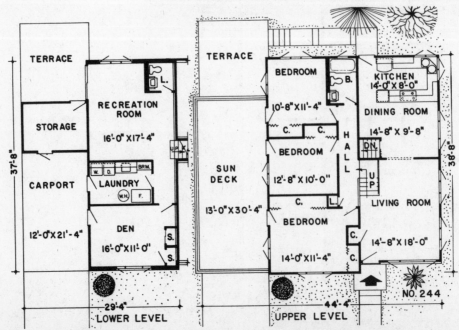

LOWER LEVEL

UPPER LEVEL

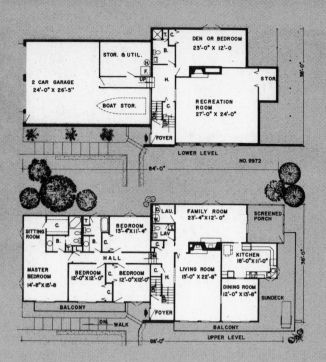

Screened porch complements balconies

No. 9972—Balconies serving the three front bedrooms and living and dining room are outdone only by the screened porch adjacent to the family room. Accessible to the large sun deck bordering the kitchen, the porch will increase the enjoyment of cool balmy breezes and provide a delightful sleeping or dining area while keeping away insects and animals. Three fireplaces add atmosphere, and five bedrooms, including a luxurious master bedroom suite with sitting room, are incorporated in the floor plan.

Upper levels—2,565 sq. ft.,
Recreation room level—1,389 sq. ft.,
Garage—1,176 sq. ft.

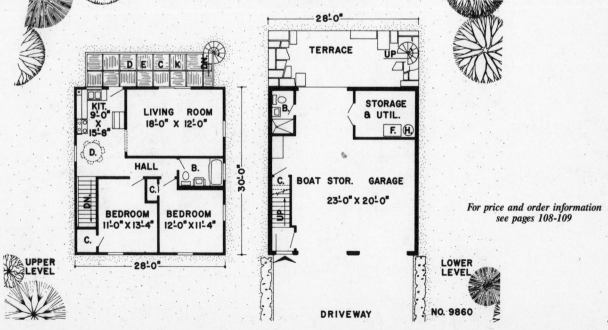

Spiral staircase highlights unique exterior

No. 9860—Leading up from a flagstone terrace to the wooden deck, the spiral staircase is a distinctive element in the unusual exterior. A mansard roof, layered with red cedar shake shingles, and the two level design allow for a great deal of space inside. Two closeted bedrooms share a full bath on the main level, and living room and kitchen both open to the wooden deck. Another half bath borders the terrace on the lower level.

Upper level—840 sq. ft., Lower level—840 sq. ft.

For price and order information see pages 108-109

71

Terrace borders kitchen, living area

No. 291—Skirting the living room and encircling the kitchen, the stone terrace affords space for picnics, sunbathing and enjoying cool evening breezes. This rugged design has an efficiently planned interior, including a kitchen complex with breakfast nook, storage space, and room for a washer and dryer. Three bedrooms, a bath and ample closet space complete the design, which would be economical to build. An additional storage area flanks the terrace to the rear of the home.

First floor—887 sq. ft.

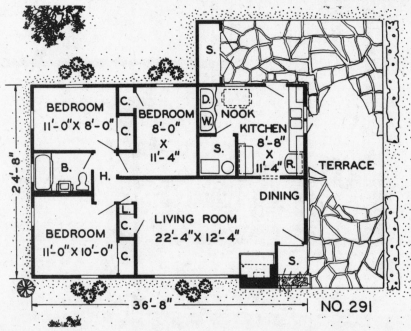

Details enhance cozy cottage

No. 298—Compartmented bath with double sinks and well-placed utility room adjoining the kitchen exemplify the careful detailing that gives personality to this three bedroom budget plan. Kitchen and dining area, open to increase the usable floor space, border the sizable living room on one side and convenient utility room on the other. Perfect for storage or laundry equipment, the utility room supplements a niche for furnace and water heater.

First floor—1,054 sq. ft.

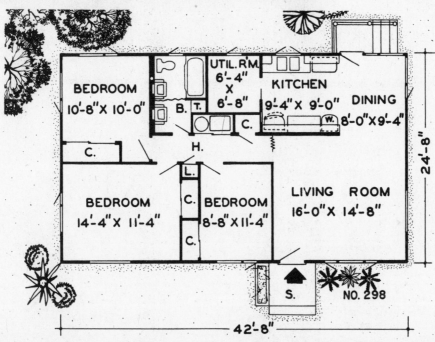

Do-it-yourself, suggests design

No. 302—Simple construction and total area of just 784 square feet make this plan a possible "do-it-yourself" project for either summer cottage or permanent home. Bedrooms are adequate both in size and closet space, and the living room spans a comfortable 17 feet and merits a coat closet. Dining room and kitchen are combined for efficiency and border a handy utility room, a logical choice for laundry center or storage area.

First floor—784 sq. ft.

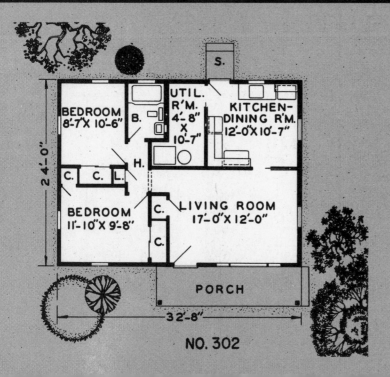

*For price and order information
see pages 108-109*

A-frame accommodating, appealing

No. 10092—Three bedrooms and a balconied 12-foot sleeping loft assure comfort for family and guests in this unique vacation home. Expanses of glass diffuse the scenery and include two pairs of sliding glass doors in living and dining rooms. Useful for picnics or restful conversation, the long decks frame both rooms, and the central wood-burning fireplace lights the two story living room. Compact kitchen flanks the dining room, and a full bath is tucked accessibly near the stairs.

First floor—1,232 sq. ft.,
Upper level—224 sq. ft.

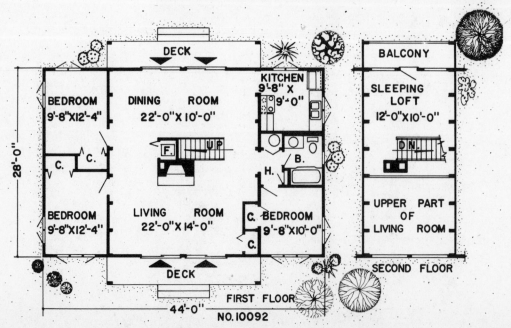

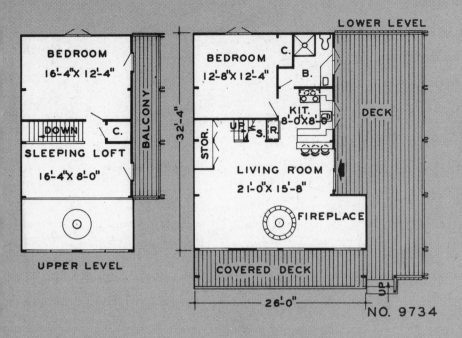

BEDROOM
16'-4" X 12'-4"

BALCONY

DOWN   C.

SLEEPING LOFT
16'-4" X 8'-0"

UPPER LEVEL

LOWER LEVEL

32'-4"

BEDROOM
12'-8" X 12'-4"

C.

B.

KIT.
8'-0" X 8'-0"

DECK

STOR.   UP   S.R.

LIVING ROOM
21'-0" X 15'-8"

FIREPLACE

COVERED DECK

26'-0"

NO. 9734

A-Frame enjoys view from all angles

No. 9734—Upper balcony, large wooden deck and expanses of windows in this modified A-frame permit maximum enjoyment of the natural surroundings . . . whether on a wooded hillside or along the seashore. Two large bedrooms, plus a sleeping loft that would easily accommodate bunk beds, translate into a great deal of sleeping space. The spacious living room with fireplace opens onto the covered deck, and the design also allows for a small, well-arranged kitchen and adequate closet space.

**Main floor—727 sq. ft.,
Upper level—406 sq. ft**

Expansive living room encourages enjoyment

No. 9760—Open planning in the living room places a dining area next to the kitchen and leaves a large uncluttered area, radiating from the wood-burning fireplace, for pure enjoyment. Relaxed informality is encouraged by the efficient breakfast nook in the kitchen and a balcony bordering the upstairs sleeping loft. The private bedroom on the main level neighbors its own full bath. Doors on both ends of the home will allow cross-ventilation of fresh mountain or ocean breezes.

**First floor—896 sq. ft.,
Upper level—199 sq. ft.**

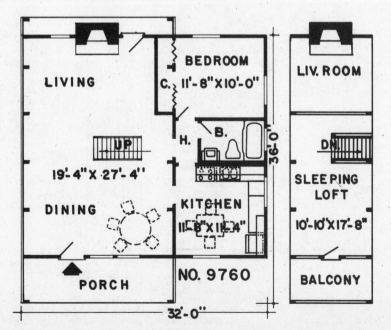

LIVING

BEDROOM
C. 11'-8" X 10'-0"

LIV. ROOM

UP

H. B.

19'-4" X 27'-4"

36'-0"

DN

SLEEPING LOFT

10'-10" X 17'-8"

DINING

KITCHEN
11'-8" X 11'-4"

PORCH   NO. 9760   BALCONY

32'-0"

*For price and order information
see pages 108-109*

Private patio, pool season summer

No. 9666—Devised for undisturbed enjoyment of the outdoors, the patio and connecting pool area of this sprawling ranch are completely encircled in decorative fencing. The patio meets the family room through sliding glass doors, and a central hall reaches the fireside living room. A full bath with shower serves both the master bedroom and pool, and another bath serves the two remaining bedrooms and family room. Endowed with light from corner windows, the kitchen overlooks the partially shaded patio.

First floor—1,409 sq. ft., Basement—663 sq. ft., Garage—565 sq. ft.

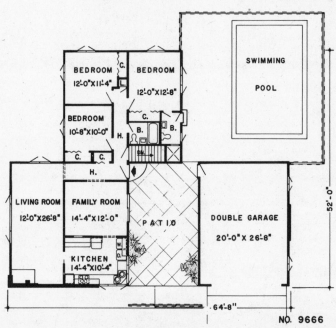

Less than 1,200 square feet

No. 9680—An entrance foyer and center hall provides access to all rooms. Roman brick, grooved plywood siding and a solar block wall are skillfully blended together to provide a very attractive exterior. The covered patio will serve as an outdoor dining area and the sliding windows in the kitchen provide a handy food pass-through. A bath and a half are on the main floor and a bath with shower is in the basement. The storage room behind the carport provides storage space for items used outdoors.

**First floor—1,196 sq. ft.,
Basement—1,196 sq. ft., Carport—336 sq. ft.**

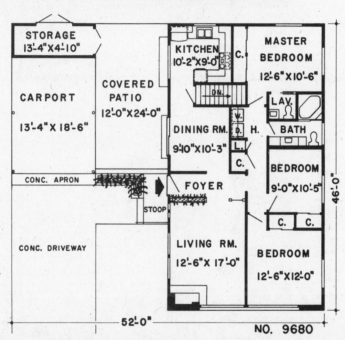

STORAGE 13'-4"X4'-10"

CARPORT 13'-4" X 18'-6"

COVERED PATIO 12'-0"X24'-0"

KITCHEN 10'-2"X9'-0"

MASTER BEDROOM 12'-6"X10'-6"

LAV.

DINING RM. 9'-10"X10'-3"

BATH

BEDROOM 9'-0"X10'-5"

CONC. APRON

FOYER

STOOP

CONC. DRIVEWAY

LIVING RM. 12'-6" X 17'-0"

BEDROOM 12'-6"X12'-0"

46'-0"

52'-0"

NO. 9680

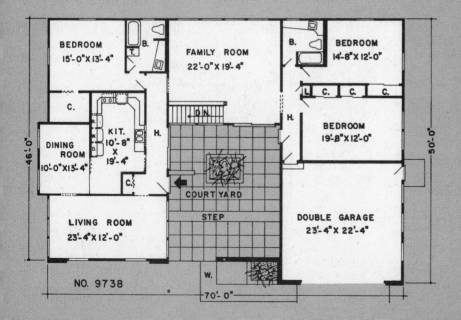

BEDROOM 15'-0"X13'-4"

FAMILY ROOM 22'-0"X19'-4"

BEDROOM 14'-8"X12'-0"

KIT. 10'-8" X 19'-4"

DINING ROOM 10'-0"X13'-4"

BEDROOM 19'-8"X12'-0"

COURTYARD

STEP

DOUBLE GARAGE 23'-4"X22'-4"

LIVING ROOM 23'-4"X12'-0"

46'-0"

50'-0"

NO. 9738

70'-0"

Home encircles exotic courtyard

No. 9738—Central to this rambling contemporary home is the sunken courtyard, which yields private outdoor relaxation and a profusion of natural light through sliding glass doors to the bordering hall, living room and family room. A formal dining room protrudes outward slightly and suggests an enjoyable view while dining. The large kitchen will lodge washer and dryer and even an informal eating area if desired.

**First floor—2,237 sq. ft.,
Basement—1,554 sq. ft.,
Garage—552 sq. ft.**

*For price and order information
see pages 108-109*

Airy atmosphere pervades split level

No. 9048—Down three steps from the gracious foyer, the main level of this well-ordered split level breathes with light and space. The lavish array of windows brightens the living room and recreation room beneath, both of which are endowed with wood-burning fireplaces. The master bedroom is enriched with its own full bath and built-in dressing table and towel closet, while the main bathroom encompasses both tub and shower. An open air porch is reached via sliding glass doors from the living room and dining room.

**First floor—1,763 sq. ft.,
Basement—1,763 sq. ft., Garage—459 sq. ft.**

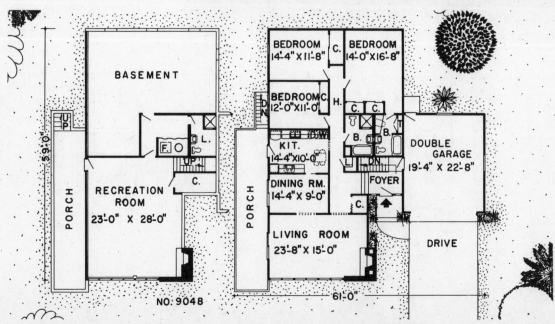

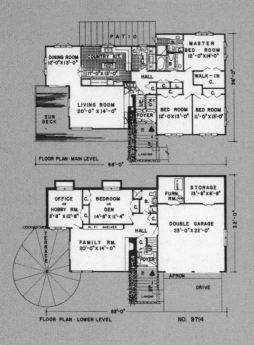

FLOOR PLAN · MAIN LEVEL
66'-0"

FLOOR PLAN · LOWER LEVEL
62'-0"
NO. 9714

Plan excellent choice for sloping lot

No. 9714—Utilizing a sloping lot to create a striking design, this split foyer plan embodies outdoor living areas and highly livable lower level. Front-facing and opening to terrace, the family room dominates the lower level, which also includes bedroom, hobby room, and full bath with shower. Above a sun deck greets the living room and dining room, and a tiled country kitchen, complete with cooking island and built-in laundry, and three bedrooms and two full baths comprise the sleeping wing.

**Main level—1,748 sq. ft.,
Lower level—932 sq. ft., Garage—768 sq. ft.**

Formal and informal living

No. 9596—If you long for the exciting atmosphere of contemporary styling, this may be the design for you. Plank and beam roof construction has been used throughout the house. The living room is sunken below the main floor level, providing a luxurious appearance. The family-dining room is the nucleous of the house, channeling traffic to all areas. Extra large closets and two full baths are provided. The balcony is reached through sliding glass doors in both end bedrooms. The open stairway in the family room leads to the very large recreation room on the lower level. The covered patio and large terrace will add to the pleasures of outdoor living.

First floor—1,876 sq. ft., Basement—818 sq. ft., Carport—434 sq. ft.

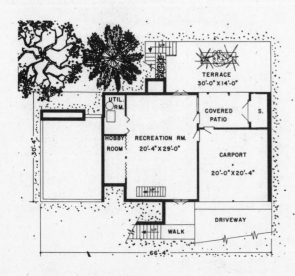

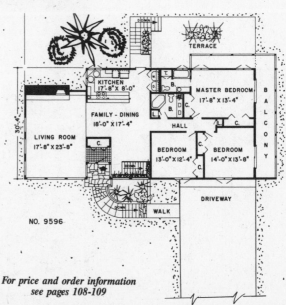

NO. 9596

*For price and order information
see pages 108-109*

Contemporary design
involves outdoors

No. 9862—Bedrooms are particularly favored in this contemporary plan, which supplies an angled balcony for the master bedroom and patios for each of the smaller bedrooms, open via sliding glass doors. In addition, the L-shaped living and dining room have access to a side terrace, while an enclosed court enhances the front. Full bath with shower serves both bedroom area and family room with fireplace, and laundry and eating space are provided in the functional kitchen. Built-in china cabinet benefits dining area.

First floor—2,155 sq. ft.,
Basement—2,155 sq. ft.

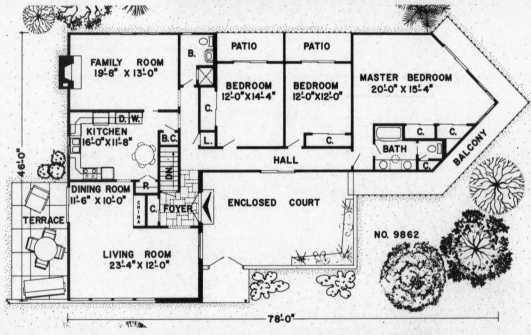

FAMILY ROOM 19'-8" X 13'-0"

B.

PATIO

PATIO

MASTER BEDROOM 20'-0" X 15'-4"

BEDROOM 12'-0"X14'-4"

BEDROOM 12'-0"X12'-0"

KITCHEN 16'-0"X11'-8"

D.W.

B.C.

C.

L.

C.

BATH

C. C.

C.

BALCONY

HALL

46'-0"

DINING ROOM 11'-6" X 10'-0"

P.

CHINA C. FOYER

ENCLOSED COURT

NO. 9862

TERRACE

LIVING ROOM 23'-4" X 12'-0"

78'-0"

J. R. BRESLER

Balcony watchtower to three winds

No. 9896—Encircling the unique living room and family room, the expansive balcony distinguishes this split foyer design and enjoys breezes from three directions. The living room with fireplace melts into the family room and country kitchen. Four bedrooms occupy the sleeping wing and include a master bedroom suite, with dressing room, walk-in closet and compartmented bath. The lower level houses a bedroom and bath and a huge recreation room that spills onto a sunken terrace.

**Upper level—2,269 sq. ft.,
Lower level—1,482 sq. ft., Garage—806 sq. ft.**

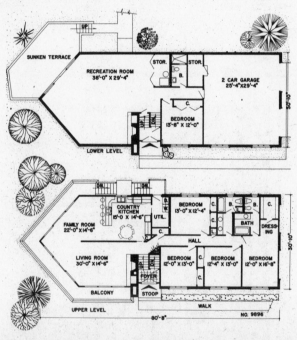

*For price and order information
see pages 108-109*

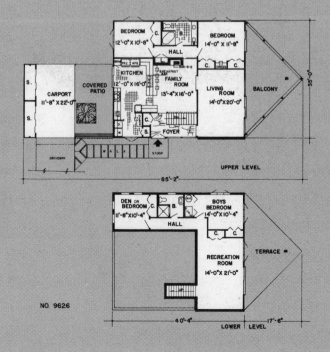

Master bedroom enjoys
V-shaped balcony

No. 9626—Designed to absorb the natural beauty surrounding it, the futuristic V-shaped balcony proves a practical asset to this contemporary layout. Two bedrooms and a bath are shown on both levels, in addition to a lower recreation room edging the roofed terrace. On the upper level, a bedroom and the living room open to the balcony, while the elongated kitchen skirts the covered patio, perfect for casual dining. The kitchen itself is intricately detailed, featuring built-in desk, a laundry niche and breakfast bar, and the adjacent family room is favored with an indoor barbecue grill.

**First floor—1,446, Lower level—660 sq. ft.
Carport and storage—348 sq. ft.**

Raised Ranch has popular features

No. 9694—Has a split foyer entrance and a large sun deck. The sun deck is accessible from the living room through sliding glass doors or from the terrace via a stairway. Three bedrooms and two full baths are on the main level and a huge family room, bedroom and bath with shower are on the lower level. An outside stairway permits direct access to the outdoors from the utility room. The garage is extra deep, providing a large storage area at the rear.

**Upper level—1,471 sq. ft.,
Lower level—1,113 sq. ft., Garage—622 sq. ft.**

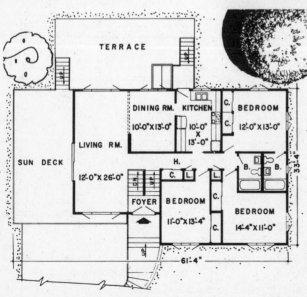

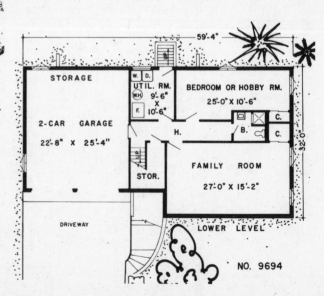

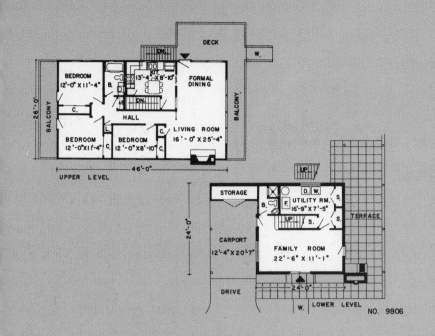

Take advantage of the view

No. 9806—This compact but very attractive home is intended for a sharply sloping lot. A lake shore site intended for year-around living would be quite appropriate. The balconies provide outdoor living space and add spaciousness to the exterior appearance. Sliding glass doors are used quite extensively to provide an abundance of natural light and ventilation. An efficient floor plan utilizes both levels to provide maximum use of all available space. A vented barbeque grill on the covered terrace provides a smokeless outdoor cooking area.

Upper level—1,196 sq. ft.,
Lower level—576 sq. ft., Carport—295 sq. ft.

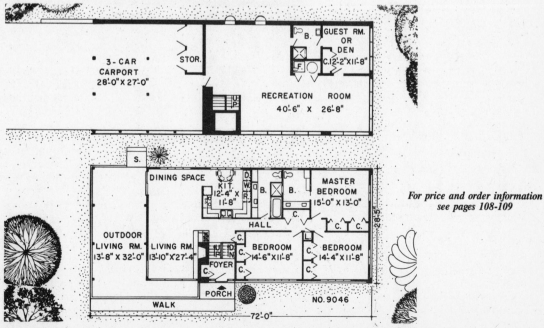

Outdoor living room awaits summer

No. 9046—Relaxing outdoor barbecues and fun-filled parties will find a natural setting in the inviting outdoor dining room of this bi-level design. Inside, an immense recreation room provides an alternative and is bordered by a den and full bath. Three bedrooms are found on the upper level, including a master bedroom with lavish private bath with shower. The elongated living room features a fireplace and dining area and skirts a sizable kitchen with laundry and eating space.

First floor—1,624 sq. ft.,
Basement—1,130 sq. ft., Carport—784 sq. ft.

For price and order information see pages 108-109

The Product Information Source

A wealth of Building Product Information available to you free or at nominal cost. Use the order form on pages 95 & 96 to obtain the literature you want.

Appliances

IN A WORLD OF CLAMOR, COMES A QUIETER DISHWASHER. New dishwashers feature significantly reduced operating sound levels and give consumers the freedom to match the appliance to any kitchen decor. New line includes five built-in and two convertible dishwashers. All models feature three-level wash action and the exclusive Micro-Mesh filter. *Maytag Co. Circle No. 255.*

KITCHEN UNIT WITH MICRO-WAVE OVEN. For the first time, a microwave oven is available with a 30" kitchen unit in one factory-assembled unit. Materials describe further details. *King Refrigerator Corp. Price 25¢. Circle No. 486.*

GROUND FAULTS. Handy guide on how to avoid ground faults in the home that can injure or kill. Describes how ground faults occur and what can be done to make the home safer. New ground fault circuit interrupters eliminate hazard of an appliance grounding and injuring a child or adult. *General Electric. Circle No. 253.*

RANGE HOODS. 12-page full color brochure describes Broan Mfg. complete line of kitchen range hoods including the Supreme 86000 — a heavy-duty range hood designed specifically for convertible cooktops with barbeque grilles. Hoods are available in all popular appliance colors and stainless steel. *Broan Mfg. Circle No. 283.*

MICROMATE. Off the counter and over the range, the Broan MicroMate combination range hood and microwave shelf puts your microwave where it belongs — over the range at eye level. Full color brochure explains how MicroMate can save valuable counter space and provide quiet, high-performance kitchen ventilation. *Broan Mfg. Circle No. 284.*

THERMADOR/WASTE KING. Leading manufacturing of Microwave Ovens, Micro-Convection Ovens, Micro-Thermal Ovens, Self-Cleaning Ovens, Range Hoods, Drop-In Ranges, Electric and Gas Cooktops, Hoodless Cooktops, Indoor Gas Barbeques, Warming Drawers, Dishwashers, Disposers, Trash Compactors, Can Openers, Kitchen and Bathroom Ventilators, Built-in and Portable Room heaters. *Thermador/Waste King. Circle No. 285.*

Bathroom

BATH CABINETS. Color catalog shows array of charming bath cabinets and ensembles. Decorator models include cabinet equipped with digital clock and night light. Stained-glass model has matching mirror. Bath lighting includes new valances. Winged mirror models in many sizes and finishes. *Nutone. Price $1.00. Circle No. 532.*

BATHE YOURSELF IN QUIET ELEGANCE. No longer is the bathroom the forgotten room in your house. We've created an atmosphere of total comfort and elegance, a pleasant place just for you . . . with softly expressed lines that are pleasing to the eye and with delicate curves that are molded for the contours of your body. Coupled with this grace and comfort is amazing practicality. *Acrylic Tubs Inc. Price 50¢. Circle No. 408.*

EVERYTHING FOR THE BATH. Full-color brochure illustrates full line of lavatories, bathtubs, water closets & whirlpools. You'll find exactly what you're looking for, with color and style to compliment your home. *Briggs, a Jim Walter Co. Circle No. 104.*

SAUNA PLANS AND EQUIPMENT SELECTION. Full-color brochure with plans and instructions for installing authentic Finnish saunas. Modular and precut, sauna rooms in sizes from 4'x4' to 8'x12'. Choose one of the standard plans or follow recommendations for designing your own custom sauna. A complete line of quality heaters, controls and accessories. *Amerec Corp. Price 75¢. Circle No. 409.*

BATH ACCESSORIES. Full-color folder illustrates Accentware bath accessories, as practical as they are beautiful. Includes dual-track shower bar which lets you dress your tub enclosure with draperies. Switch plates, etc., for use in any room. *Kirsch Co. Circle No. 191.*

ACCESSORIES. Full color catalog of decorative HallMack bathroom accessories . . . towel bars, towel rings, garment hooks, tissue holders, soap dishes. Complete selection of brass and solid oak fixtures plus a unique assortment of built-in accessories to accent any decor. *NuTone. Price $1.00 Circle No. 547.*

ELEGANT BATH. "Kohler Elegance" offers 44 pages of colorful ideas for bathrooms, powder rooms and kitchens; whirlpool bath, spas, lavatories, kitchen and bar sinks; water-saving toilets, shower heads and faucets. Booklet helps with product selection, color coordination and decorating ideas. *Kohler Co. Price $2.00. Circle No. 546.*

BATH FANS AND HEATERS. 12-page, full-color catalog describes the full line of Broan bath fans and auxiliary heaters including the Model 655 combination heater/fan/light. Model 655 offers the convenience of instant bathroom warmth, efficient ventilation and bright lighting, all in one unit. *Broan Mfg. Circle No. 286.*

WHIRLPOOL BATHS. Full-color catalogue will show you the largest product line in the industry. You can choose from more sizes, shapes, colors and combinations of units than anywhere else. *Jacuzzi® Whirlpool Bath. Circle No. 107.*

TILE SHOWER WATERPROOFING. Asphalt membrane specially designed for waterproofing under tile shower floors. Easy to install and economical. Protects subfloors from water damage from your shower. *Compotite Shower Pan. Circle No. 109.*

ACRYLIC ELEGANCE FROM ATI & MORE. Description of optional steam bath and Jacuzzi whirlpool options on acrylic bathtubs featuring exclusive "Odyssey" bubble door. *Acrylic Tubs Inc. Price 50¢. Circle No. 545.*

Built-ins

THE MASTER SPACE-SAVER. Murphy Beds help to create multi-purpose rooms by providing sleeping accommodations as comfortable as the best standard bed which can also be concealed simply and safely with a minimum of effort. They are not a compromise like a sofa bed and can even be put away fully made up and ready to use. They are easy to conceal because they use a simple and efficient counterbalancing mechanism proven by years of field experience. *Murphy Door Bed Co. Price 50¢. Circle No. 534.*

BUILT-IN IRONING CENTERS. Iron-A-Way by any standard of excellence is the finest ironing center ever built. It has become a household word throughout the United States and provides a compact pressing facility without any storage preparation or putting-a-way problems. When the door is closed, everything is hidden, and it can be painted to match the room or faced with a full-view mirror or other decorative elements. *Iron-A-Way. Circle No. 218.*

CENTRAL VACUUM SYSTEM. Built-in cleaning power right at your fingertips. Eliminates wasted motions while adding to the value of your home. Full color brochure describes the extra cleaning power of the Central Vac and easy installation. *Broan Mfg. Circle No. 287.*

CENTRAL CLEANING SYSTEM IN YOUR HOME. New 32 page booklet is off the press. It provides all you need to know about installation and maintenance of a central vacuum cleaning system. Fully illustrated with drawings and photographs. *Wal-Vac Inc. Price 50¢. Circle No. 549.*

BUILT-IN FOOD CENTER. For the up-to-date kitchen, built-in Food Center operates 10 separate appliances including Food Processor and new Coffee Grinder. Motor is installed under the counter to eliminate kitchen clutter. Full-color catalog describes operation of each appliance in detail. *NuTone. Price 50¢. Circle No. 548.*

Doors & Windows

HOW TO INSTALL INTERIOR JAMBS AND EXTERIOR DOOR FRAMES. 8-page brochure for do-it-yourselfers explaining the differences between interior jambs (both flat and adjustable) and exterior frames, planning requirements, tools and supplies needed, step-by-step installation procedure illustrations, plus much more. *Wood Moulding & Millwork Producers. Price 40¢. Circle No. 491.*

DOORS OF A LIFETIME. Lifetime Doors manufactures doors of all types including flush hollow or solid core, raised panel, white pine louver/panel and more in finished or unfinished. Color brochure describes and shows all types available. *LifeTime Doors. Circle No. 116.*

ENTRY SYSTEM. For new construction or replacement, the Therma-Tru system represents excellence in design and manufacture. Full line, 20-page color brochure illustrates design features and benefits of product line which includes deeply embossed or raised panel designs, French Patio systems, Crystalline Series™ leaded, beveled glass styles, and B Label Fire Rated systems. *Therma-Tru. Circle No. 290.*

ENJOY YOUR HOME MORE WITH A BILCO BASEMENT DOOR. When building your new home, be sure it has this key to a convenient, useful, safe basement . . . a modern, all-steel Bilco Door. It supplements your interior stair. Eliminates unnecessary tracking through first floor rooms. Makes storage easy, convenient. Send for free folder with sizes, construction data and names of local dealers. *The Bilco Co. Circle No. 121.*

JIM WALTER DOORS. Residential and commercial doors, including sectional garage doors, rolling and side coiling doors and grilles, fire doors, door operators and commercial hollow metal doors. *Jim Walter Doors. Circle No. 291.*

CUSTOM SHUTTERS, BLINDS. Ask for our illustrated brochure. *Ohline Corp. Circle No. 289.*

INTERNATIONAL DOORS. Brochure describes International Doors from the finest vertical grain Douglas fir or Western Hemlock. Simpson uses original art in the designer series to create solid wood doors with accents. Traditional doors are also featured, and special care is always taken to provide elegance at a realistic price. *Simpson Timber Co. Price 75¢. Circle No. 420.*

PLASTIC-VIEW. Heat reflective See-Thru window shades provide outstanding all weather window insulation, plus excellent protection against glare and sun damage to interior furnishings. Plastic-View shades also create daytime privacy whereby insiders can see out but outsiders can't see in . . . the security of a one-way mirror. *Plastic-View Inc. Circle No. 125.*

WORLD'S LARGEST DOOR CATALOG. Unique 68 page full-color catalog packed with over 1000 hard-to-find quality millwork items. Entry doors, French, sash and interior doors, bifolds, patio doors, energy saving door systems, screen doors, spindles, stair parts, columns, posts. Of special note: leaded glass and laser carved doors. *E. A. Nord Co. Price $2.50. Circle No. 510.*

WINDOW AND GLIDING DOOR REPLACEMENT. 20-page, full-color folder is devoted exclusively to replacing existing windows and gliding patio doors. Answers most often asked questions about window replacement. Illustrates and describes different styles of low-maintenance Andersen® Perma-Shield® windows and gliding doors available for replacement. Shows energy savings possible with new windows. *Andersen Corp. Circle No. 288.*

TROCAL RIGID VINYL WINDOWS. 6 page brochure describes double-hung residential windows which combine the energy saving advantages and ruggedness of Trocal commercial windows with the flexibility required for new residential installations or remodeling. *Trocal Window Systems, Div. Dynamit Nobel of America Inc. Circle No. 113.*

GUIDE TO ENERGY-SAVING WINDOWS. 16 page booklet includes the components that comprise a quality window; various styles of windows, proper window locations for optimum light, ventilation, and fuel savings; condensation and the effect quality windows can have on energy usage. *National Woodwork Mfg. Assoc. Price 50¢. Circle No. 550.*

Fireplaces & Stoves

WOODMASTER PRIMER. Descriptive pamphlet showing savings on home heating costs by burning wood as opposed to oil and electricity, with charts showing the heating values of wood and savings in dollars on wood versus oil/electricity. *Suburban Mfg. Co. Circle No. 259.*

ULTIMATE IN FIREPLACE EFFICIENCY. Colorful brochure describes the operation of fireplace inserts. Provides illustrated details of air flow, installation, and dimensions, and features the WOODMASTER Fireplace Insert and Deluxe Fireplace Insert along with specifications of each. *Suburban Mfg. Co. Circle No. 261.*

WOOD SPLITTING MADE EASIER. Literature tells of a unique axe which splits many woods in a single stroke. Chopper 1™ possesses rotating levers which transform the downward stroke to a powerful outward force and also prevents the axe from sticking in the wood making the splitting of firewood faster and easier than ever before. *Chopper Industries. Circle No. 128.*

EASY LOG LIGHTING. Blue Flame log lighting system makes heating with modern pre-fab fireplaces more energy efficient by using less than 1¢ of natural gas to quickly ignite a roaring log or coal blaze. System is easily installed in either pre-fab or existing masonry fireplaces. Saves oil, gas and electricity in home heating. *Canterbury Enterprises. Circle No. 133.*

FIREPLACE INSERTS. Thermograte's new economical fireplace insert will beautify and convert your fireplace into

an efficient heat producer for less than $400. The unit insures safe operation and conserves heat with dual tempered glass doors. The adjustable damper, stainless steel box, dual speed blower, and extended warranty provide incredible value at this low price. *Thermograte Inc. Price $1.00. Circle No. 544.*

WOOD BURNING FIREPLACES. Full-color performance report and do-it-yourself installation planner brochures feature manufacturer's most efficient fireplace, the Warm Majic, rated 41-43% efficient by the Wood Heating Alliance. Fireplace is available with outside air and glass enclosure kits. Brochures include color photos and illustrations of the fireplace and chimney system installations. *Majestic. Price 50¢. Circle No. 552.*

DISTINCTIVE FIREPLACES BY RICHARD LE DROFF. Full-color brochure illustrates 18 self-contained imported fireplaces from France. Many unique, rustic, and formal styles included. Also, complete masonry facades for existing or zero-clearance fireplaces. Famed European authority offers suggestions on fireplace selection. *Adams Co. Price $1.00. Circle No. 551.*

WOOD MANTELS GIVE OLD FIREPLACES NEW LIFE. Hand-crafted wood mantels offer the professional builder, remodeler or homeowner-handyman, an economical and easy way to add new life and looks to any fireplace. *Readybuilt. Circle No. 227.*

FIREPLACES, STOVES, & INSERTS. Freestanding models in black or porcelain colors. Some models are glass enclosed for clear view of fire. All models have energy saving features. *Malm Fireplaces Inc. Circle No. 293.*

REBEL FIREPLACE INSERT OR FREE-STANDING STOVE. Burns wood slowly and more completely. Reburns captured heated gases for maximum utilization of energy. Heats up to 3,200 sq. ft. with as little as 10-degree local temperature variation. Made of heavy-gauge steel triple wall construction, double wall damper, cast iron doors, blower system, refractory lined firebed, matte black finish and solid brass hardware. Available in four sizes with many deluxe options. 10 year warranty. *Hutch Mfg. Circle No. 292.*

Floors & Walls

BEAUTIFUL WALLS & FLOORS. Full color brochure highlights the entire Z-BRICK product line, including decorative facing brick and stone, flooring products, RUFF-IT Acrylic Sculpture Coat and All-Weather Stucco. *Z-BRICK Co. Circle No. 280.*

HOW TO WATERPROOF YOUR HOME. A fold-out brochure features schematic drawings of the exterior of a house and an inside basement wall with concrete and masonry trouble spots pinpointed. It includes a special section on how to correct damp basement walls. *Thoro System Prod. Price 10¢. Circle No. 515.*

REDWOOD PANELING. Available in a wide choice of patterns, grades, sizes and textures, redwood offers a rich natural complement to walls, ceilings, and accent areas. It's a natural insulator against both heat and cold. *Simpson Timber Co. Price 50¢. Circle No. 513.*

NEW LOOK. Brochure describes how you can give your home a New Look with lustrous hardwood parquet floors. There is an extra special warmth and cordial feeling you get with parquet floors. Your parquet floors will be a conversation piece and the envy of those around you. *Peace Flooring Co., Inc. Circle No. 137.*

CARPET FOUNDATION. Densified prime urethane foam used under wall to wall carpeting and also under your most prized area rugs. Material includes two brochures. One explains the benefits of OMALON and the other is a practical Room Planner/Decorating Kit to help you plan your room settings around your lifestyle. *Olin Corp. Circle No. 264.*

PROFESSIONAL OAK FLOOR THAT YOU CAN INSTALL YOURSELF. Booklet describes how easily Hartco flooring can be installed. Color photos show the shades available in different flooring finishes. Describes Hartco's exclusive Par-K-Stik with foam back as well as the plain back or foam back types. Also includes floor care product information. *Hartco. Circle No. 294.*

NATURESCAPES. Brochure offers an exciting design alternative for residential or commercial decor. Naturescapes photomurals are durable, dry-strippable, and meet all institutional standards. *Naturescapes, Inc. Price $1.00. Circle No. 431.*

ALTERNATIVE FLOORING. Brochure illustrates the use of ceramic floor tile. Tile that is durable enough for floors and yet light enough to use with coordinating walls. *Marazzi USA Inc. Circle No. 139.*

TESSITURA II. 25 real fabric wallcoverings in 87 colorways on the exclusive "Tessitura" ground. 4 companion fabrics in 15 colorways also available. *James Seeman Studios, Div. of Masonite Corp. Circle No. 262.*

THIS GOOD EARTH II. Collection of wallcoverings and companion fabrics inspired by natural elements from all over the world. 26 wallcovering designs in 100 colorways are shown with the 8 fabrics in 34 colorways. *James Seeman Studios, Div. of Masonite Corp. Circle No. 263.*

THE TOP FLOOR SINCE 1898. Parquet and plank flooring. Full-color booklet illustrates a full line of parquet block and plank flooring. The many patterns come unfinished and prefinished. Booklet includes specification chart. *Harris Mfg. Co. Price $1.00. Circle No. 516.*

THE PANELING BOOK. 28 page booklet includes how-to instructions, diagrams, and full-color photos of rooms you'll want to duplicate. Make this your complete guide to paneling any room, whether building or remodeling your home. *Georgia-Pacific. Price $1.00. Circle No. 429.*

DECORATIVE STONE. Brochure describes manufactured building stone veneer made from lightweight concrete. Although available at a fraction of the cost of natural stone, it is hard to distinguish from the real thing. The light weight permits easy installation over nearly any existing or new surface. *Eldorado Stone Corp. Circle No. 138.*

PLAQUES, TABLEAUX, & TILES. Set of 3 booklets illustrating the Authentic OLD DUTCH HAND PAINTED DELFT tiles, tableaux and ceramic relief plaques. The individual tiles are stocked in blue crackled and non-crackled, and POLYCHROME crackled, as well as tableaux in POLYCHROME crackled. *Amsterdam Corp. Price $2.50. Circle No. 517.*

HOW TO WATERPROOF MASONRY WALLS. 16 page booklet illustrates the common causes of seepage and damp basement walls and shows techniques and products for treating interior and exterior masonry surfaces to withstand moisture penetration. From the manufacturer of DRYLOK Masonry Sealer. *United Gilsonite Laboratories. Price 25¢. Circle No. 557.*

HARDWOOD FLOORING. Ten-page, full-color brochure describes and illustrates a line of pre-finished hardwood flooring. Seven popular flooring patterns are shown in various room settings. Diagrams of each pattern are included with pattern size specifications. *Robbins Inc. Price 25¢. Circle No. 556.*

WOOD FLOOR CARE PRODUCTS, FACTORY-FINISHED MOLDINGS & ADHESIVES. Brochure describes Hartco floor care products and chart shows which product is recommended for a particular finish. Color photos show the factory-finished moldings to match Hartco floors. Included is information on latex adhesives available to install the wood flooring. *Hartco. Circle No. 295.*

MEMPHIS HARDWOOD FLOORING. Full-color 12 page catalog of pre-finished and unfinished Chickasaw floor styles shows solid oak plank, parquet and strip in gorgeous designer-inspired room settings. *Memphis Hardwood Flooring. Price 25¢. Circle No. 555.*

ULTIMATE IN HARDWOOD FLOORING. Brochure presents the complete product line, from the custom classics and end grain floors to basic plank and parquet. Line is directed toward the upper-end specifier and consumer for that special residential accent area. *Kentucky Wood Floors Inc. Price $2.00. Circle No. 554.*

ACCENT ON PANELING. Emphasizes simple but effective solutions to decorating problems with just 1, 2, or 3 sheets of paneling. Shows how to use paneling to build-in seating or storage, create dramatic feature walls, enlarge small spaces or enhance furniture projects. *Georgia-Pacific. Price 50¢. Circle No. 553.*

GLORIA VANDERBILT. Collection of wallcoverings & companion fabrics designed by Gloria Vanderbilt. 21 wallcovering designs in 104 colorways are shown with the 12 fabrics in 52 colorways, plus sections on "The Border Story," "Tulip Time in Town & Country," "A rainbow of ideas you can adapt," "Dress Up a Closet," etc. *James Seeman Studios. Circle No. 296.*

Furniture & Furnishing

ELEGANT SHEEPSKIN & LAMBSKIN RUGS. This color brochure describes the beauty, warmth and durability of a variety of luxuriously soft wool sheep and lambskin rugs, the ultimate in dynamic decorating. A special offer on these exceptional New Zealand woolskins is made possible through a limited shipment. New Zealand produces some of the highest quality wool anywhere due to the moderate climate. In most wool producing regions of the world, the cold winters will produce brittle areas in the wool fiber, causing these fibers to gradually break and make the fleece thinner. This doesn't happen to New Zealand wool. New Zealand care has converted one of nature's most versatile resources into fine prestigious furnishing rugs to add distinction and elegance to any home. Find out how you too can own one of these most delightful decorator pieces. *Treasures. Circle No. 210.*

HOW TO BUILD A CEDAR CLOSET. Brochure describes and illustrates three different ideas for solving your storage problems. Included are plans for a free-standing cedar storage closet, an attic closet, and an existing closet. *Home Closet Planning Service. Circle No. 297.*

SPACE BUILDER. Ventilated shelving by Closet Maid is the modern method of solving storage space problems quickly and easily. Creates additional storage space without remodeling in bedroom closets. Can turn a blank wall into a pantry in kitchens. Keeps cleaning supplies high and dry in utility rooms. *Closet Maid. Circle No. 234.*

OLD SOUTH REPRODUCTIONS OFFERS A RETURN TO ELEGANCE. Classic and timely styling from bygone days, masterfully recreated for today. Cast aluminum tables, chairs, settees, coat rack, umbrella stands . . . hundreds of singularly distinctive pieces for home and patio. Full-color catalog depicts wide range of items in lovely finishes. *Moultrie Mfg. Co. Price $1.00. Circle No. 460.*

CLASSIC CHAIRS . . . AND MORE. 16-page, full color catalog illustrating a large variety of classic imported chairs and tables, both contemporary and traditional, of wood, chrome, leather, wicker, marble and glass. *Intrends Int., Div. of Walker & Zanger, Inc. Price $2.00. Circle No. 495.*

FURNITURE. Now you've finished your home you want furniture to complete the picture. Catalog shows the complete line of furnishings for every room in your home. *Terra Furniture Inc. Price $5.00. Circle No. 559.*

WROUGHT IRON-INDOORS & OUT. Color brochure features traditional and contemporary lines of very comfortable casual furniture available in a variety of frame colors, woven vinyls and designer fabrics. Choose from dining and seating groups, tables, chaises or servers. Baker's rack, corner shelves, etageres, plant stands and other matching accessories also available. *Lyon-Shaw. Price $2.00. Circle No. 558.*

Hardware

GARAGE DOORS & OPENERS. Frantz has the best of both, shown in full color in two new catalogs. Model homes and driveway designs help in home planning. Opener is state-of-the-art model — new programmable design with top safety features. *Frantz Mfg. Co. Circle No. 282.*

DISTINCTIVE TOUCH OF CLASS. Catalog of unusual, hard-to-find hardware, plumbing, and lighting fixtures. Quality reproductions are made from solid brass, porcelain, oak and wrought iron. Complement 18th and 19th century decor as well as contemporary homes.

Renovator's Supply, Inc. Price $2.00. Circle No. 519.

DECORATIVE HARDWARE. 32 page decorating booklet shows how to select and install hardware items to coordinate each room with matching style door locks, bath hardware, cabinet hardware, and wall switch plates. Several designer hardware styles are detailed. *Amerock Corp. Price 25¢. Circle No. 433.*

HANDCRAFTED HARDWARE OF THE COLONIES. For over a third of a century, Acorn has been reproducing the charm and warmth of Early American Hardware. Booklet shows full line of antique looking hardware for your home. Beautiful yet sturdy and practical. *Acorn Manufacturing Co., Inc. Price $3.00. Circle No. 504.*

MIDGET LOUVERS. Louvers for venting air, light, heat, sound and moisture. Brochure describes method of prevention and correction of "moisture blistering" of house paint and "dry-rot" fungus control. Easy to install directions. *Midget Louver. Circle No. 299.*

RELIABLE SIMPLICITY IN DOOR CONTROLS. Dorma Door Controls, Inc., has published a unique 16-page product guide to READING series surface applied hydraulic door controls. The comprehensive catalog features an easy-to-read guide to ANSI numbers, and has a cross index to comparable competitive products. In addition, there is a Typical Application section that lists doors by application and the correct surface applied hydraulic door control for each particular application. *Dorma Door Controls, Inc. Circle 298.*

Heating & Cooling

AIR CLEANING. Booklet describes air cleaning and presents efficiency data on common household pollutants and how to control them with a new type air cleaner. *Research Products Corp. Circle No. 152.*

KEEP IT MOVING. Lomanco describes how to keep air moving with their complete line of ventilation products. Includes whole-house central fans, turbine ventilators, ridge line ventilators and power ventilators. *Lomanco, Inc. Circle No. 231.*

HUMIDIFICATION. Booklet discusses relative humidity and lists several points to consider when purchasing a humidifier. *Research Products Corp. Circle No. 151.*

HOW TO GET THE MOST FROM YOUR COMFORT INVESTMENT. 16 page booklet explains the importance of clean air, humidity and air circulation to your comfort system. Includes sections on solar heating and efficient two-speed cooling. Plus tips on saving energy. *Lennox Industries, Inc. Circle No. 229.*

CARRIER'S WEATHERMASTER III. Literature describes the only heat pump system on the market which houses the compressor and all critical controls out of the weather. Compressor and controls are installed in garage, utility room or basement, meaning improved system reliability as critical components are protected from rain, snow and freezing temperatures. Concept, system operation, features, accessories, physical data and dimensions, and performance data all included. *Carrier Corp. Circle No. 194.*

ENERGY SAVINGS WITH NATURAL VENTILATION. Bulletin describes how whole house ventilation system can be used for summer comfort and energy savings by reducing air conditioning. Details advantages and describes the solid state speed control and timer control features. Installs in any home. *Kool-O-Matic Corp. Price 25¢. Circle No. 435.*

AIR MOVER GRILLE. Moves warm or cool air where it is needed. The economical alternative to revamping a central heating or cooling system which fails to maintain uniform temperatures in adjoining rooms or between floors. *Hutch Mfg. Circle No. 281.*

AIR CIRCULATOR GRILLE. Redistributes air to make temperature more uniform. Utilizes electric fan to draw warm air or cool air from one room and redirect it into another level of the same room or to an adjoining room. Saves money on utility bills and operates on mere pennies a day. *Hutch Mfg. Circle No. 265.*

ROOM-TO-ROOM FAN. New Room-to-Room Fan is ideal for distributing warm air heated by fireplace or wood-burning stove to adjacent rooms. Cata-

log offers complete selection of exhaust fans for use in bathrooms, laundry rooms and kitchens. *NuTone. Price 50¢. Circle No. 560.*

CARRIER'S DELUXE ROUND ONE. Brochure describes a sophisticated central air conditioning system that combines high efficiency operation with a moderate price. Included is a detailed analysis — in layman's language — of the Deluxe Round One's quality construction. *Carrier Corp. Circle No. 302.*

WHOLE HOUSE VENTILATOR. Full color brochure explains the easy, no-joist cutting installation of the Broan Whole House Ventilator and both the direct drive and belt drive models. Complete line of accessories including shutter, timer, speed control and thermostat is also featured. *Broan Mfg. Circle No. 301.*

BROAN COLLECTOR SERIES CEILING FANS. Ultra-quiet operation and elegant styling. Brochure describes exclusive design, transistorized speed control, beautifully hand-crafted housing finishes, and full line of accessories. Available in both 42" and 52" models. *Broan Mfg. Circle No. 300.*

THE ENERGY SAVERS. Tabloid describes summer related energy saving products including air conditioner and cooling air deflectors. *Deflecto Corp. Circle No. 149.*

EASY INSTALLATION. Brochures describe whole house attic fans that are designed to cool a house at great energy savings and are a breeze to install by the average do-it-yourselfer. *Emerson Environmental Prod. Circle No. 232.*

Home Plan Books

DUPLEX, CONDOMINIUM AND MULTI-UNIT HOME PLANS. The Garlinghouse Company has consolidated their multi-unit designs into one source. You'll find over fifty designs suitable for residential or resort settings including ranch, Tudor, Swiss and Spanish designs. Includes duplexes, triplexes, fourplexes and larger. *Garlinghouse. Price $3.50. Circle No. 530.*

SINGLE-LEVEL AND UNDERGROUND HOME PLANS. This book from the Garlinghouse Company provides you with over 130 new home plan designs to aid you in choosing and planning the perfect home. Energy efficient underground and bermed homes are two of the special features in this book. Low-

The Product Information Source

cost construction drawings are available for all designs. *Garlinghouse. Price $3.50. Circle No. 538.*

MULTI-LEVEL AND SOLAR HOME PLANS. The latest in energy-saving designs and solar use are just two of the many unique features in this new book of over 120 home plan designs (many in full color). You are sure to find the home to fit your individual preferences in this comprehensive collection of truly great designs. *Garlinghouse. Price $3.50. Circle No. 539.*

SMALL HOME PLANS. Modest in square footage and economical to build, but designed with comfortable family living in mind, is the emphasis of this publication. This 116 page book contains all styles from traditional to contemporary, single to split-level, sleek to rustic. Many solar designs with energy efficient features. *Garlinghouse. Price $3.50. Circle No. 584.*

TRADITIONAL HOME PLANS. Over 125 traditional designs, many with solar energy features, are found in this book from the Garlinghouse Company. English Tudor, Cape Cod, Victorian, Oriental and Classic Farmhouse are just several of the styles. Many full-color illustrations and easy to read floor plans. *Garlinghouse. Price $3.50. Circle No. 531.*

COMPLETE HOME PLANS COLLECTION. This 10-book library comes complete with a handsome plastic binder. Packed with more than 750 different home designs (many in full color), this collection will be the one and only source you'll need to find that special home. There are hundreds of designs of all sizes, styles, and configurations imaginable, and low-cost construction drawings with energy-saving features are available for each one. If you are serious about a new home, then this collection is a must. *Garlinghouse. Price $19.50. Circle No. 505.*

Kitchens & Cabinets

SOLID COLOR COLLECTION. Illustrating the 64 top decorative colors of the decorative laminate. *Wilsonart. Price 50¢. Circle No. 502.*

CABINETS. Beautifully designed contemporary door styles accent today's modern kitchens. Birchcraft's fine assortment of stylish patterns allows you to choose the perfect door to fit your personal tastes as well as the wood and stain finish that best compliments your overall design. Ranging from sleek, ultra-modern to country casual, the Birchcraft line of custom cabinet styles is sure to compliment the most epicurean tastes. *Birchcraft Kitchens. Circle No. 154.*

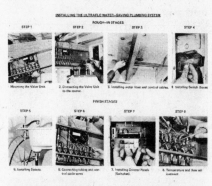

PUSH BUTTON PLUMBING. Booklet details Ultraflo, a system which conserves both water and energy and can be installed quickly at much less cost than conventional plumbing. Reduces the water bill up to 35% and saves about 25% on the heating of water. *Ultraflo Corp. Circle No. 110.*

KITCHEN SINKS. Complete catalog gives specifications and photos of a complete line of stainless kitchen sinks. *Polar Ware Co. Price 25¢. Circle No. 403.*

KITCHEN & BATH PLANNING GUIDE. This 16-page color booklet contains helpful suggestions on how homeowners can get the "dream kitchen" or bath of their choice. It covers subjects such as efficient "work triangles," where to get professional help, and a consumer guide to cabinet selection. *National Kitchen Cabinet Assoc. Price 50¢. Circle No. 540.*

LIVING CENTER. Distinctive kitchens designed for the living patterns of your family are shown in full-color booklet. Planning procedures to use space wisely for well-organized step saving work areas are discussed thoroughly. *Kitchen Kompact Inc. Price $3.00. Circle No. 445.*

ELKAY BOOKLET OF SINK/FAUCET COMBINATIONS. Elkay stainless steel sinks and other products give a smart contemporary look that flatters any interior, blends with every color scheme. Work-saving accessories in many models and sizes. Easy cleaning . . . longer wear! *Elkay Mfg. Co. Price 50¢. Circle No. 522.*

THE CENTER OF FAMILY LIVING. New four color packet, while showing cabinet styles of oak, pine, pecan and maple, provides necessary and helpful information on kitchen design and layout as well as decorating ideas. Matching bathroom cabinetry, hutches and wall systems provide added storage ideas for every room of your home. *Arist-O-Kraft. Price $1.00. Circle No. 444.*

PEEL & STICK LAMINATE. Peel & stick self adhesive laminate for do-it-yourself countertops, desks, table tops, etc. *Conolite. Circle No. 303.*

SPECIFICATIONS AND ACCESSORIES GUIDE. The literature contains diagrams and dimensions for all units, and shows Haas' wide array of accessories and specialty cabinets. *Haas Cabinet Co. Circle No. 304.*

CREATIVE CABINETRY. Brochure gives illustrations, components, dimensions and cabinet construction for creative cabinetry for every room in your home from kitchen, family room and laundry room to bedrooms, entertainment areas and powder rooms. *Connor Forest Ind. Price 50¢. Circle No. 561.*

COUNTRY MAPLE STYLING. Four-page color brochure features Rich Maid kitchen designs equipped with cabinetry and various exclusives. Also included are samples of how doors and drawers can be treated with antique accents. *Rich Maid Kitchens, Inc. Price 35¢. Circle No. 562.*

OFFICIAL KITCHEN & BATH COLOR & DESIGN GUIDE. Six kitchens and four bath designs by Wilsonart, Tarket Gafstar, Magic Chef and Kohler, adaptable for use in new homes or remodeling projects are seen in this 64-page, full color guide. *McKone & Co., Inc. Price $4.00. Circle No. 563.*

Lighting

A TOUCH OF CLASS. Leviton's Decora line of rocket switches, available in all home harmonizing colors are indeed a decorative touch in a home that adds "class." Dimmers both of the wall variety and table top are also important effects. Even dimmers that are touch sensitive, has a memory and has a true decorator look. *Leviton Manufacturing Co., Inc. Circle No. 162.*

ARCHITECTURAL LANTERNS. Trimble House architectural lanterns feature colonial, victorian and contemporary styling to harmonize perfectly with buildings and landscape new and old. All lanterns are totally cast in aluminum offering maintenance free, functional construction with enduring elegance of design. *Trimblehouse Corp. Circle No. 161.*

HOME CONTROL SYSTEM. From Leviton, a modular remote control plug in system which permits the user to control up to 16 lights or appliances from one location in addition to convenience a dimming feature provides energy savings. A single button for all lights, off or on, also has the advantage of security. Starter kit includes control box and 2 modules. *Leviton Mfg. Co., Inc. Circle No. 213.*

SKYLIGHTS. New catalog offers the largest line of residential and light commercial skylights. Included are: Daylighter roof window made of a solid hardwood frame and insulated glass; Daylighter Long-Lites which fit between the beams of a house; and the complete line of Sky-Vue fixed and ventilating skylights. *APC Corp. Circle No. 305.*

Moulding and Millwork

FYPON ENTRANCE SYSTEMS. Brochure features molded millwork for distinctive entrance ways. Moldings, window features specialty millwork, door entrance systems, bow and bay window and roof and shutters. *Fypon Inc. Circle No. 118.*

VICTORIAN MOULDINGS AND MILLWORK. Unique "turn of the century" mouldings and millwork is available for today's houses. Includes mouldings, headblocks, baseblocks, casings, wainscoat and baseboard in a standard stock of premium grade pine and oak. Redwood, mahogany and other woods available on custom basis. Catalog illustrates products as well as construction drawings showing the typical uses. *Silverton Victorian Millworks. Price $3.50. Circle No. 509.*

HOW TO FINISH WOOD MOULD-INGS. Shows how to apply a professional looking finish. Explains differences between paints, stains, clear finishes and miscellaneous finishing products. Illustrates how to finish new wood. Covers tools needed and cleanup. *Wood Moulding & Millwork Prod. Price 40¢. Circle No. 568.*

DESIGN & DECORATE. Contains full-color photos of various household rooms showing extensive use of wood mouldings. Profile drawings of all

mouldings used to create these finished designs are shown, plus suggested alternative profiles. Simple ways to transform an ordinary room into a showplace on a minimum budget. *Wood Moulding & Millwork Prod. Price 75¢. Circle No. 567.*

HOW TO WORK WITH WOOD MOULDINGS. For do-it-yourselfers, brochure gives step-by-step basics of working with wood mouldings, including measuring, mitering, coping and splicing. Shows most standard profiles available plus ideas on uses. *Wood Moulding & Millwork Prod. Price 40¢. Circle No. 566.*

THE COLLECTION. Beautiful, pre-engineered architectural ornamentation in lightweight modern materials. Products include mouldings, ceiling medallions, mantels, overdoor pieces, niches, domes, stair brackets, etc. Pre-engineered and primed for easy installation. Great for restoration or remodeling. All products can be painted or stained. *Focal Point. Price $2.50. Circle No. 565.*

VICTORIAN RESTORATIONS. 24 page catalog illustrates dozens of solid wood Victorian millwork designs based

upon turn-of-the-century originals. Entire line made from kiln dried, premium grade hardwoods. Includes millwork for exterior and interior use. Customizing is available. *Cumberland Woodcraft Co., Inc. Price $3.50. Circle No. 564.*

Patio & Yard

COMPLETE OUTSIDE BASEMENT ENTRANCE! You probably know the sales appeal direct basement access gives your home. Perm-Entry makes it easy and economical for you to include it in your home. The concrete stairwell is manufactured and installed to rigid specifications, and capped with the rugged watertight heavy gauge steel PermEntry door. *Perm Entry Co. Circle No. 122.*

ORNAMENTAL GATES & FENCES. Color catalog illustrates a selection of ornamental gates and panels for fencing. Master plan illustrates how to design a brick wall containing ornamental fence panels. Catalog also includes complete line of outdoor furniture for poolside, fountains, urns, planters, garden pla-

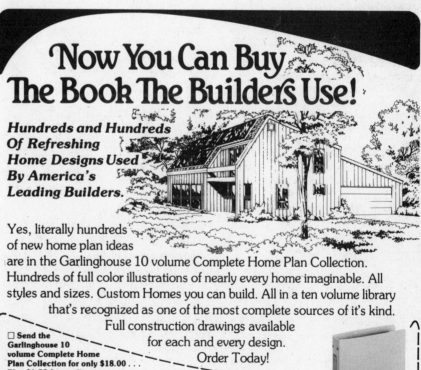

ques and mailboxes as well as a variety of small gift items. *Moultrie Mfg. Co. Price $1.00. Circle No. 447.*

GARDEN SETTINGS. New eight-page color booklet, Redwood Garden Settings for Spas, Tubs, Pools, features spectacular idea-starting hot tub surrounds, poolside decks and spa shelters, all using the economical redwood garden grades. Design and planning tips can help you build a luxury redwood retreat in your own backyard. *California Redwood Assoc. Price 50¢. Circle No. 475.*

SPATUB. The "new" Spatub looks identical to a hot tub from the exterior, but resembles a spa on the interior with molded seats and flexshell liner. Comes ready to assemble including rim deck and roll top. *California Cooperage. Price $1.00. Circle No. 569.*

STORAGE BUILDINGS. A complete line of galvanized steel and aluminum storage buildings. The styles include gable, high gable, gambrel and high gambrel; various sizes and colors. Are factory cut and drilled. *Arrow Group Industries. Circle No. 306.*

DECKS ADD FUN & VALUE. This year enjoy your home inside and out by investing in contemporary outdoor living. Redwood Design-a-Deck Plans Kit can help you get started. As with any project, careful advance planning assures more efficient deck construction. This kit helps minimize confusion by giving you everything you need to know, including materials lists, information on which redwood grades to use and where, which finishes to apply and how. Nails and fastening systems, as well as pre-cut deck patterns, 20-page instruction manual and planning grid. *California Redwood Assoc. Price $5.00. Circle No. 541.*

WHIRLPOOL SPAS. Full-color catalogue shows the full range of spas and accessories available with detailed description design. *Jacuzzi® Whirlpool Bath. Circle No. 307.*

SHOWERS OF DIAMONDS. Fountains by Rain Jet. Sculptured patterns from rotating fountain head provides arresting beauty. Alive! Vital! Fascinating! Lights and color blender available for jewel-like effects. Water recirculates. Complete fountain bowls to 8 feet. Full color catalog. *Rain Jet Corp. Circle No. 308.*

UNDERGROUND SPRINKLERS. Brochure shows how to install Rain Jet permanent underground sprinkler system with a minimum of sprinkler heads and digging. Patented nozzles distribute rain-like droplets evenly. Choice of heads includes squares, strips, circles, etc. Flexible pipe and full flow fittings make installation easy. *Rain Jet Corporation. Circle No. 309.*

SPAS. 16-page color brochure describes complete line of fiberglass and acrylic spas in both custom in-ground and portable models. Come totally preplumbed for easy installation and are designed and built for years of hassle-free operation. *California Cooperate. Price $1.00. Circle No. 570.*

REDWOOD, HOT WATER & YOU: A PERFECT MIXTURE. A colorful brochure describing the joys of owning your own hot tub. Gordon & Grant redwood hot tubs are individually hand-crafted by highly skilled coopers from carefully selected clear all-heart kiln-dried vertical grain redwood. Every tub is an individual. *Gordon & Grant. Price $1.00. Circle No. 571.*

DECKS & OUTDOOR PROJECTS. Includes how to plan your deck, what materials you'll need, the best building techniques, plans for an economy deck, terraced decks, and a gridwork deck, plus planters, tables and a buffet cart. *Georgia-Pacific. Price $1.00. Circle No. 572.*

DECORATIVE CREATIONS. Fiber-

glass garden pools, redwood waterwheels, fiberglass "rock" waterfalls, fountains & lights, bubbling fantasias, garden bridges, pumps & accessories to create tranquility, serenity and beauty in your garden setting. *Hermitage Gardens. Price $1.00. Circle No. 573.*

Roofing

HOW TO DO IT . . . AND SAVE. Color brochure featuring economy grade red cedar shingles and Handsplit shakes. Filled with ideas on the use of #2, #3 and #4 shingles plus #2 shakes on walls, planters, furniture and in many other ways around the home. *Red Cedar Shingle & Handsplit Shake Bureau. Price 10¢. Circle No. 449.*

OVER-ROOF WITH CEDAR. Color brochure showing how home owners can save applying a beautiful cedar shingle or shake roof right over their old roof. Detailed "how to" information plus data on tools and techniques. *Red Cedar Shingle & Handsplit Shake Bureau. Price 10¢. Circle No. 448.*

REROOFING GUIDE. Step-by-step instructions for the do-it-yourself reroofer. Includes formula for estimating shingle coverage, tips for roof inspection and preparation, methods for stripping an old roof, shingle application, nailing schedules and flashing applications. *Georgia-Pacific. Price 25¢. Circle No. 574.*

A HOMEOWNER'S GUIDE TO QUALITY ROOFING. When to replace an existing roof, should you do the job yourself, how to pick a contractor and a review of general roofing procedures are discussed in this brochure. Also discusses the features of asphalt shingles and illustrates the latest styles and colors. Contains a color guide to help coordinate roofing with siding and trim. Plus a section on talking the roofer's language and tips on the care of roofs. *Asphalt Roofing Mfg. Price 50¢. Circle No. 575.*

Security

KEYLESS ENTRY. Pushbutton combination door locks. Press a four digit combination from the outside and door unlocks automatically. 10,000 possible combinations. Mechanically activated, no electricity. Interior unlocks by pressing a button. With built-in nitelatch, 1" throw deadbolt defies forcible entry. No keys to lose or misplace. Guaranteed pickproof! *Preso-Matic Lock Co. Circle No. 177.*

SUPER SECURITY. Latch-Gard II, a security device for homes, apartments, motels and hotels, offers real security. A door can be opened safely one inch to view the would-be intruder. The super tough aircraft cable cannot be violated as easily as the common chain door guard. Brochure describes use & proper installation instructions. *Latch-Gard, Inc. Price 25¢. Circle No. 543.*

Siding & Insulating

REDWOOD SIDING. 4 color brochure offers a variety of handsome richly-toned siding grades and patterns to complete the redwood exterior with a natural complementary touch. Redwood is one of the most durable softwoods, famed for its ability to withstand weather. *Simpson Timber Co. Price 50¢. Circle No. 528.*

WALLBANGER'S POCKET GUIDE. Step-by-step instructions on exterior wall application of cedar shakes or shingles for the do-it-yourselfer, or the professional. "How to" on tools, equipment, estimating and application. *Red Cedar Shingle & Handsplit Shake Bureau. Price 25¢. Circle No. 576.*

REDWOOD INTERIOR/EXTERIOR GUIDE. Highlights top-quality exterior siding and interior paneling in 8 pages. Applications and techniques for building with the many grades and patterns of redwood are illustrated. Solid information for any specifier or user. *California Redwood Association. Price 50¢. Circle No. 577.*

SAVE MONEY BY INSULATING. Booklet offers instructions and advice on how and where to save money by insulating, weatherstripping and caulking the home. For the do-it-yourselfer, safety measures, proper tools and simple instructions with illustrations. Tips on selecting a contractor and dealing with him in his own language. Where to go to find out about costs and how to take advantage of a tax credit are also covered. *Mineral Insulation Manufacturers Association. Price 50¢. Circle No. 578.*

ATTIC INSULATION. Literature provides step-by-step instructions for the do-it-yourselfer on how to install fiber glass or rock wool insulation in the attic floor. Everything from how to calculate the proper amount of insulation to how to provide for adequate ventilation is

covered. There's even a section on how to hire a contractor if that's preferred. Plus a section on R-values. *Mineral Insulation Manufacturers Association. Price 45¢. Circle No. 583.*

Solar

SOLAR SHINGLE. Full color brochure describes the solar shingle, a unique approach to the solar collector. The shingles provide a dual service as solar collectors and as a roofing material which can be used to heat your pool, spa or hot tub. They are aesthetically pleasing while maintaining thermal efficiency. *Jet Air, Inc. Circle No. 239.*

SOLAR COLLECTORS. Features flat plate copper collectors for use in domestic hot water and space heating. Designed to be efficient, yet economical while having an anticipated life span of 20 years or more. *National Solar Supply. Circle No. 241.*

HEATING. Booklet titled "Solar Energy Facts" contains background information, detailed descriptions of all current types of solar heating units, as well as their practical applications. A special point of emphasis is given to the economies of solar heating in today's economy. *Research Products Corp. Circle No. 268.*

V-12 SOLAR COLLECTOR. Heat With The Sun! Save our vital resources while you save on your monthly utility bills. "Daytime Assist" and "Complete—With Storage" solar systems available. Brochure describes solar collectors that heat your home and water. *Sunflower Energy Works, Inc. Circle No. 240.*

SOLARMATE HOT WATER SYSTEMS. 4-page brochure describes the quality line of efficient and dependable solar hot water systems. *Lennox Ind., Inc. Circle No. 242.*

TEMPERATURE MONITOR LOCATES IN LIVINGROOM. To show off and monitor energy saving systems or solar installations. Recess mounts install into standard wall box. Sensors can be located as far from monitor as necessary. *Q Systems. Circle No. 310.*

AMETEK POWER SYSTEMS DIVISION. Currently offering its Sunjammer Solar collectors to qualified contractors and builders for installation in residential space and hot water heating systems. The product incorporates an aluminum extruded housing finished with the

Sherwin-Williams PowerClad® baked enamel paint finish, high transmission solar glazing, and all copper absorber plate using black chrome selective coating. 20, 26, and 33 square feet sizes available. *Ametek Power Systems Division. Circle No. 188.*

Specialties

BEAUTIFUL WOOD FINISHING. 12-page, 4-color "How to Beautifully Finish Wood" booklet enables anyone to create professional finishes the first time with all types of wood. One application of Watco Danish Oil seals, primes, finishes, hardens, protects, beautifies, penetrates deeply into the wood; outlasts surface coats three to one. Nine attractive alcohol-base stain colors also available. *Watco-Dennis Corp. Circle No. 270.*

STAIR LOCK. Literature describes a stairway system called "Stair Lock" which features pre-machined interlocking parts which can be assembled and installed in one hour, with no technical skill required. Reversible parts create a variable rise and run that is designed to fit most of today's homes. *Visador Co. 35¢. Circle No. 501.*

RADIO-INTERCOM. Unique radio intercommunication systems for homes. Telephone may be answered from any speaker installed in home, garage or patio. Luxury systems to economical packaged radio intercoms provide safety in door answering. *Nutone. Price $1.00. Circle No. 526.*

RESIDENTIAL INSTALLATION GUIDE. A brochure featuring installation details and performance data is available for Residential TJI® joists, MICRO=LAM® headers, and MICRO=LAM beams. Longer and lighter than conventional solid sawn joists, TJI joists are capable of up to 24" o.c. spacings. MICRO=LAM laminated veneer lumber headers and beams are stiffer and stronger than solid timbers and beams. *Trus Joist Corp. Circle No. 314.*

The Product Information Source

ALUMINUM GREENHOUSES. Prefabricated aluminum and glass for a plant growing environment, active and passive solar use, hot tub covers, pool covers, etc. Over a hundred models to choose from. Custom fabricate for locations where none of the standard models will work. *Aluminum Greenhouses Inc. Price $2.00. Circle No. 579.*

HOMEBUYER'S CHECKLIST. The Checklist is an easy to use guide to insure that no item which is important is overlooked or forgotten when a family is building or buying a home. The checklist contains over 200 items concerning the home, the lot, the neighborhood and the community. *General Home Service. Price $1.00. Circle No. 580.*

WOOD MAILBOXES. 8-page brochure for those wanting the perfect decorator mailbox for the front of their home. Describes several distinctly different handcrafted wood mailboxes in rural and house mounted styles along with posts, planter bases and paper tubes. May be personalized with names and numbers carved into the wood surfaces. Plans & kits also available. *Spielmans Wood Works. Price $1.00. Circle No. 581.*

THE FINISHING TOUCH. Color illustrated 16 page booklet with tips for easy wood finishing and refinishing. Provides step-by-step instructions for all phases of a typical wood refinishing process — from removing old paint to the selection and application of new stain and protective wood finishes. From the manufacturer of ZAR wood finishing products. *United Gilsonite Laboratories. Price 25¢. Circle No. 582.*

IN-THE-FLOOR-SAFE. A revolutionary, low cost, In-The-Floor Safe has been developed by Meilink Safe Company. This new unit comes in a key lock version or a key and combination version. Capacity is 1188 cubic inches. *Meilink Safe Co. Circle No. 279.*

STATELY COLUMNS & ORNAMENTAL GRILLE WORK. Columns in all sizes, fluted or smooth, round or square, in any length, stand stately and carefree in restoration projects or new construction. Color catalog shows other aluminum products; cast railings, columns, grilles, fences and gates in many elegant designs from the past. *Moultrie Mfg. Co. Price $1.00. Circle No. 542.*

PURE DRINKING WATER. General Ecology's SEAGULL IV residential drinking water purification device is the modern replacement for bottled water. SEAGULL IV removes a wide variety of contaminants including bacteria, TCE, chlorine, and asbestos fibers, making ordinary tap water taste fresh and delicious. Selected for 1982 World's Fair, Energy Efficient Home. *General Ecology, Inc. Circle No. 246.*

WATER FILTER. Literature describes remarkable unit that removes sediment and dirt from all household water. Model VIH installs easily on the main cold water line. Valve is built into the head of the filter. The unit has three modes of operation: on, off and bypass, permitting quick, easy change of cartridges. Fabricated completely of durable, corrosion free, high impact molded plastic. *Filterite Corp. Circle No. 172.*

UNDERSTANDING UNDERGROUND WATER. This booklet describes the benefits of individual water systems, discusses types of wells and water system components, and tells how to size a system. *Water Systems Council. Price 45¢. Circle No. 455.*

ORDER YOUR WATER WELL DONE. A guide to locating and constructing water wells. Water quality control is also discussed. *Water Systems Council. Price 75¢. Circle No. 454.*

ELIMINATE TANGLED CORDS. Literature describes the new Chopper Cord Keeper. This unique storage device will hold up to 100 feet of electrical extension cord, rope, and flat nylon hose. This product is a real benefit for homeowners. *Chopper Industries. Circle No. 312.*

ENERGY SAVING MONEY SAVING ITEMS. Tabloid tells of energy saving products designed to cut down on high utility bills. Everything from air deflec-

tors, water heater blankets, clothes dryer heat deflectors to pipe insulation. *Deflect-o Corporation. Circle 245.*

DOOR CHIMES. Brochure describes the complete line of Broan Door Chimes including the Model 977 Selectune Programmable Chime. Model 977 plays 24 melodies and is an imaginative way to greet guests. Brochure also describes complete line of push buttons and non-electric chimes. *Broan Mfg. Circle No. 311.*

PLANNING A HOME LAUNDRY CENTER. Booklet discusses design, location, storage, work space and possible pitfalls to be avoided. *Maytag Co. Circle No. 247.*

CHIMNEY FLUE CAP. All metal welded construction. Keeps out rain and snow, preventing rust damage to dampers. Keeps in sparks and cinders. Stops birds and squirrels from entering chimney. Stops leaves and other debris from collecting in the flue. *Hutch Mfg. Circle No. 273.*

MAYTAG ENCYCLOPEDIA OF HOME LAUNDRY. Contains material on energy conservation, fibers, fabric finishes, stain and spot removal and information about home laundry planning and features available on laundry equipment. *Maytag Co. Price $1.25. Circle No. 525.*

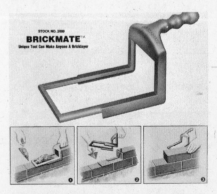

BRICKMATE. A unique tool that maintains the correct amount of mortar consistently between each brick and each course of bricks plus keeps proper alignment at the same time. *Roy Enterprises. Circle No. 313.*

Literature order form

Order the Information You Want Today . . .

STEP I. Circle the numbers corresponding to the literature in this publication that you want. Please enclose $1.50 processing fee.

FREE BROCHURES & CATALOGS:

104	107	109	110	113	116	118	121
122	125	128	133	137	138	139	149
151	152	154	161	162	172	177	188
191	194	210	213	218	227	229	231
232	234	239	240	241	242	245	246
247	253	255	259	261	262	263	264
265	268	270	273	279	280	281	282
283	284	285	286	287	288	289	290
291	292	293	294	295	296	297	298
299	300	301	302	303	304	305	306
307	308	309	310	311	312	313	314

PRICED LITERATURE:

403 25¢	408 50¢	409 75¢	420 75¢	429 $1	431 $1
433 25¢	435 25¢	444 $1	445 $3	447 $1	448 10¢
449 10¢	454 75¢	455 45¢	460 $1	475 50¢	486 25¢
491 40¢	495 $2	501 35¢	502 50¢	504 $3	505 $19.50
509 $3.50	510 $2.50	513 50¢	515 10¢	516 $1	517 $2.50
519 $2	522 50¢	525 $1.25	526 $1	528 50¢	530 $3.50
531 $3.50	532 $1	534 50¢	538 $3.50	539 $3.50	540 50¢
541 $5	542 $1	543 25¢	544 $1	545 50¢	546 $2
547 $1	548 50¢	549 50¢	550 50¢	551 $1	552 50¢
553 50¢	554 $2	555 25¢	556 25¢	557 25¢	558 $2
559 $5	560 50¢	561 50¢	562 35¢	563 $4	564 $3.50
565 $2.50	566 40¢	567 75¢	568 40¢	569 $1	570 $1
571 $1	572 $1	573 $1	574 25¢	575 50¢	576 25¢
577 50¢	578 50¢	579 $2	580 $1	581 $1	582 25¢
583 45¢	584 $3.50				

Complete Order Form Next Page . . .

Come home to quality.
Come home to Andersen.™

Literature order form/continued

STEP II.   Please help us out by answering these questions:

1. Are you:
   - ☐ Building a new home
   - ☐ Buying a new home
   - ☐ Buying an older home
   - ☐ Remodeling your present home

2. If you are building a new home . . .

   A. Are you planning to start construction within the next six months?   ☐ yes ☐ no

   B. Do you presently own the land that you plan to build on?   ☐ yes ☐ no

   C. How will your home be built?
      - ☐ Through a general building contractor
      - ☐ With you acting as general contractor
      - ☐ With you doing most of the actual labor

   D. Where are you getting the construction drawings for your new home?
      - ☐ Garlinghouse or another home plans company
      - ☐ Architect or local designer
      - ☐ Provided by your building contractor
      - ☐ Other _____

3. A. Which best describes the size of the closest community or population center to where you live:
   - ☐ Under 10,000 population
   - ☐ 10,000 to 50,000 population
   - ☐ 50,000 to 100,000 population
   - ☐ 100,000 to 500,000 population
   - ☐ 500,000 to 1,000,000 population
   - ☐ over 1,000,000 population

   B. Do you live:
   - ☐ Within the above community or in its immediate suburb
   - ☐ In a rural area, some distance from the above community

4. What is your approximate age?
   - ☐ 18-24
   - ☐ 25-34
   - ☐ 35-49
   - ☐ 50-64
   - ☐ 65 or older

5. What is the age of your youngest child living at home?
   - ☐ No children at home
   - ☐ Under 3 years
   - ☐ 3-9 years
   - ☐ 10-18 years
   - ☐ over 18 years

6. What products do you plan to have in your new home?
   - ☐ Insulated (double or triple paned) windows and glass door
   - ☐ Fireplace and/or wood burning stove
   - ☐ Automatic garage door opener
   - ☐ Central Air Conditioning
   - ☐ Skylights
   - ☐ Dishwasher
   - ☐ Heat Pump
   - ☐ Microwave Oven
   - ☐ Washer and Dryer
   - ☐ Wall Paneling

7. Are you or a member of your family professionally involved in the building field? If so, please indicate how:
   - ☐ No involvement
   - ☐ Building contractor
   - ☐ Building sub-contractor
   - ☐ Architect
   - ☐ Home designer (other than architect)
   - ☐ Lumber dealer
   - ☐ Other building material supplier
   - ☐ Engineer involved in residential housing
   - ☐ Other _____
   _____
   _____
   _____

Step III.   GHPG-5

Fill in your proper mailing address:

Name _____

Address _____

City _____

State _____ Zip _____

Step IV.

Figure the amount due and enclose a check or money order

Amount due for priced literature	$	_____
Processing Fee	$	_____ 1.50
Kansas Residents Add 3½% Sales Tax	$	_____
TOTAL AMOUNT ENCLOSED	$	_____

Allow 3-6 weeks for delivery.

Step V.   Make checks payable and mail complete page to:

The Product Information Source

P.O. Box 1735, Topeka, Kansas 66601-1735

ONLY PRODUCT LITERATURE MAY BE ORDERED FROM THIS ADDRESS. TO ORDER BLUEPRINTS SEE PAGES 108-109. TO ORDER BUILDING BOOKS SEE PAGE 112.

A new look in design

No. 9962—Bountiful expanses of glass, outlined in red cedar and underscored with wooden decks, dazzle the interior of this singular home with reflections of the surrounding beauty. Each of the three bedrooms boasts abundant closet space and a balcony, and each level houses a full bath. A wood-burning fireplace lights the living room and adjoining dining area, both of which are equipped with sliding glass doors to the deck. The functional kitchen supplies a breakfast bar fringing the dining room.

**First floor—1,056 sq. ft.,
Second floor—893 sq. ft.**

*For price and order information
see pages 108-109*

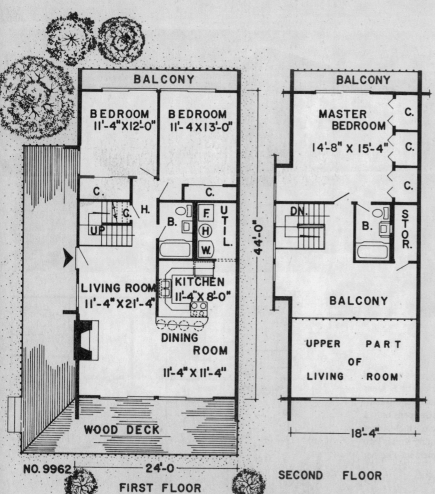

BALCONY

BEDROOM
11'-4"X12'-0"

BEDROOM
11'-4 X13'-0"

C.

C.   H.

UP

LIVING ROOM
11'-4"X21'-4"

C.

B.   F.
H
W.

UTIL.

KITCHEN
11'-4"X 8'-0"

DINING
ROOM
11'-4" X 11'-4"

WOOD DECK

NO. 9962   24'-0

FIRST FLOOR

BALCONY

MASTER
BEDROOM
14'-8" X 15'-4"

C.

C.

C.

DN.

B.

STOR.

44'-0"

BALCONY

UPPER   PART
OF
LIVING   ROOM

18'-4"

SECOND   FLOOR

Chalet can be finished as needed

No. 10026—Swiss Chalet inspired, this home allows the possibility of using it while completing the attic bedroom or lower bedroom and family room at some later date. The first floor, housing living room, kitchen, full size bedroom and bath and even a laundry would make a more than comfortable retreat until the remainder of the home could be finished. The envisioned lower level would house another bedroom or den, a large family room with fireplace, and two dressing rooms with showers and a half bath.

First floor—1,052 sq. ft.,
Second level—628 sq. ft.,
Lower level—1,052 sq. ft.

Rustic design blends into hillside

No. 10012—Naturally perfect for a woodland setting, this redwood decked home will adapt equally well to a lake or ocean setting. A car or boat garage is furnished on the lower level. Fireplaces equip both the living room and the 36-foot long family room which opens onto a shaded patio. A laundry room adjoins the open kitchen which shares the large redwood deck encircling the living and dining area. Two bedrooms and two full baths on the first floor supplement another bedroom and half bath on the lower level.

First floor—1,198 sq. ft.,
Basement—1,198 sq. ft.

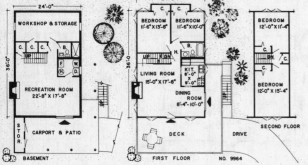

*For price and order information
see pages 108-109*

Recreation room houses fireplace

No. 9964—Restful log fires will contribute atmosphere to the sizable recreation room bounding the patio of this chalet. Upstairs, another fireplace warms the living and dining rooms which are accessible to the large wooden sun deck. Four bedrooms and two baths are outlined, and the home is completely insulated for year round convenience and contains washer and dryer space. The romantic chalet design would be equally appealing along an ocean beach or mountain stream.

**First floor—896 sq. ft.,
Second floor—457 sq. ft.
Basement—864 sq. ft.**

Home exterior blends with vacation sites

No. 10150—This two story rough frame home blends in with its surroundings near beach area or mountain retreat. Ground level entry is provided to the family room. The kitchen opens onto a living room/dining room combination, lighted by large windows and doors opening onto a three sided deck.

**Upper level—1,008 sq. ft.,
Lower level—652 sq. ft.,
Garage—356 sq. ft.**

Rustic exterior; complete home

No. 10140—Rustic though it is in appearance, the interior of this cabin is quiet, modern and comfortable. Small in overall size, it still contains three bedrooms and two baths in addition to a large, two story living room with exposed beams. As a hunting or fishing lodge or a mountain retreat, this compares well.

First floor—1,008 sq. ft.,
Second floor—281 sq. ft.,
Basement—1,008 sq. ft.

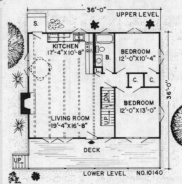

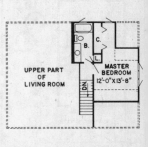

*For price and order information
see pages 108-109*

Sizable kitchen accents compact design

No. 10056—Kitchen and living room area in this all-season home comprise over half the home and encourage relaxed comfort, whether this is to be used as a vacation retreat or permanent residence. The kitchen houses a dining area and space for a washer and dryer, certainly a convenience. Two large bedrooms utilize a great deal of closet space and a hall linen closet and living room closet provide additional space. A corner fireplace and access to the wooden deck further complement the living room.

Main floor—952 sq. ft., Deck—200 sq. ft.

Rustic ranch features fireplace

No. 9076—Natural wood siding and stone chimney transport this three bedroom home deep into the wilderness, where its fireplace warms the living and dining rooms. A front porch translates a certain old-fashioned, homey comfort, but the interior claims certain modern touches, such as the master bedroom's private bath. A full bath off the hall serves the other bedrooms. Closets are plentiful, included even in the living room, and additional storage can be found in the full basement.

**First floor—1,140 sq. ft.,
Basement—1,140 sq. ft.**

Living room enjoys commanding view

No. 10024—Lounging around the living room's circular fireplace, you will delight in the panorama visible through the three pairs of sliding glass doors. Extending outward is a partially roofed redwood deck doubling as an outdoor living and dining area. Two bedrooms and compartmented bath with shower make up the sleeping area. A well -designed kitchen is furnished with a pass-through to the sizable living room. The garage leves the option of having its door in front or back, depending on your lot.

First floor—960 sq. ft., Garage—288 sq. ft.

Leisure plan poised for outdoor fun

No. 10194—Encircled by wood decks and walkways, this two bedroom vacation home opens to the outdoors on three sides for maximum enjoyment of its surroundings. Two sizable bedrooms, each favored with two closets, are separated by a compartmented bath, with the bordering laundry niche a welcome addition. Kitchen, living and dining rooms form an open area for meals and entertaining that merits an indoor barbecue grill and wood-burning fireplace. A full bath with shower is well-placed next to the deck entrance.

First floor—1,418 sq. ft.,
Parking—480 sq. ft., Outdoor storage—56 sq. ft.

Bedrooms merit access to wooden deck

No. 10220—To encourage a relaxed lifestyle and enjoyment of the outdoors, a 50-ft. wooden deck fronts this vacation retreat and opens to two bedrooms as well as the living area. Complete but simple, the plan offers a living area with two closets and a prefab fireplace, open to a compact kitchen with rear entrance. The separate laundry room also houses furnace and water heater, and the large bath features double sinks. The plan can be built without one or both bedrooms if desired.

Family area—576 sq. ft.,
Bedroom #1—168 sq. ft., Bedroom #2—144 sq. ft.

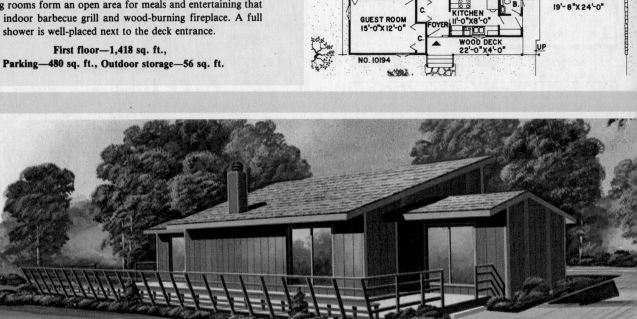

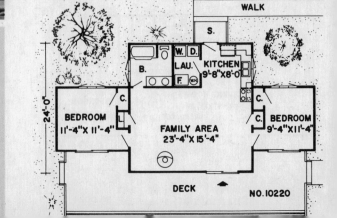

Main living area forms hexagon

No. 1052—With wood-burning fireplace creating a focal point, the main living area forms a hexagon in this contemporary retreat. The large deck, accessible from living area and both bedrooms, assures outdoor involvement, with four pairs of sliding glass doors inviting the view. Two full baths edge the bedrooms, and a totally compact kitchen is bordered by the utility/laundry room. Designed for leisure living, this plan includes two convenient storage areas.

**First floor—1,060 sq. ft.,
Outside storage—48 sq. ft., Carport—360 sq. ft.**

Deck frames master bedroom

No. 1060—Edging the master bedroom and immense living room, double triangular decks emphasize outdoor living in this creative three bedroom home. All areas of the home are heavily glassed, with windows and sliding glass doors used liberally, and have easy access to the outdoors. The living room extends over 39 feet and spotlights a central wood-burning fireplace. Other notable features of the home include a walk-in closet off the dining area, a large pantry, and a laundry room that borders the kitchen.

**First floor—2,160 sq. ft.;
Basement—1,600 sq. ft.; Carport—440 sq. ft.**

Sun deck, covered patio invite outdoor living

No. 9840—Encircling part of three sides of this home, an expansive sun deck spills off the living and dining room and allows an unparalleled view of lake or mountain surroundings. Beneath the sun deck, a stone patio balances the stone siding of the family room and is reached via sliding glass doors. The first level also includes a large hobby room, utility and storage room and half bath. Two bedrooms, a full bath, and kitchen with breakfast bar complete the upstairs plan, and a substantial sleeping loft with closet comprises the third level.

**First floor—1,120 sq. ft.,
Lower level—1,120 sq. ft.,
Upper level—340 sq. ft.**

Chalet creation sports four bedrooms

No. 9900—Besides the enticing exterior, the main concentration of this chalet design is on sleeping comfort. Four ample bedrooms, one of which opens to a balcony, endow the home with room for a number of guests. Bedrooms and hallways abound with closet space, another valuable contribution to livability. The first floor living room, dining room and kitchen are skirted by the terrace which offers outdoor dining space and is accessible from the dining room through sliding glass doors.

**First floor—936 sq. ft.,
Second floor—529 sq. ft.**

Building A-frame can be weekend project

No. 7664—Easy to apply red cedar shake shingles are specified for the roof of this A-frame cabin and help make building it yourself a feasible and rewarding weekend project. Constructed on a concrete slab, the cabin exudes relaxed informality through the warm natural tones of exposed beams and unfinished wood interior.

Main floor—560 sq. ft.; Upper level—240 sq. ft.

For price and order information see pages 108-109

A-Frame garnished with stone chimney

No. 9876—Trimmed with balcony and sun deck and garnished with a stone chimney, this A-Frame presents an engaging exterior. Inside, it is evident that the home is intended for all season use. A full bath serves each floor, including the basement which contains a huge recreation room and boat storage.

**First floor—1,232 sq. ft.,
Second floor—717 sq. ft.,
Basement—1,232 sq. ft.**

Vacation retreat enjoyable all year

No. 10170—Fully insulated ceilings and outside walls and first class construction will assure a leisure home that can be enjoyed throughout the year in this space-conserving two bedroom plan. On the lower level, a full bath with shower adjoins the laundry/utility room, and garage and boat storage are provided. Simple but efficient, the upstairs floor plan outlines two closeted bedrooms and full bath with linen closet. The kitchen is open to the spacious living/dining area, and sliding glass doors open to the deck for sunbathing or savoring the view.

Upper level—860 sq. ft.,
Lower level—836 sq. ft.

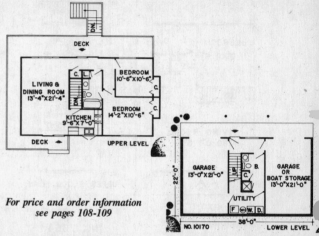

*For price and order information
see pages 108-109*

Triple-winged home angled for outdoors

No. 10096—Stretching out to capture sunshine and view, this plan employs two wooden decks and completely zoned areas to create an outstanding vacation retreat. Two Dutch doors adorn bedrooms and open them to elongated deck. Bedrooms are separated by a full bath and grouped for privacy. Permeated with the illusion of light and space induced by the cathedral ceilings and gable end windows, the sizable living room wing opens to the front deck via sliding glass doors. Kitchen and family room are joined to form a functional center for dining and laundry chores, while exposed beams and access to deck emphasize relaxation.

First floor—873 sq. ft., Storage—30 sq. ft.

Double-winged plan absorbs scenery

No. 10282—Expanses of glass, especially in living and family rooms, open this two bedroom design to its surroundings. Accessible and attractive, wooden decks encircle the plan, making it an ideal choice for shore or woodland setting. To the right of the closeted entry, the living room stretches over 19 feet to offer space for relaxed conversation and a view. The kitchen is compact and borders a family-dining room, open to the adjoining deck. Two sizable bedrooms are separated by a compartmented full bath.

**First floor—1,136 sq. ft.,
Basement expansion room—387 sq. ft.**

Conversation pit marks unique plan

No. 10288—Distinctively shaped like a sunflower, this unusual plan radiates from a central conversation pit off the family room. An open living arrangement calls for a kitchen equipped with snack bar and sandwiched between dining room and family room, a spacious area lined with sliding glass doors to the patio. Three bedrooms are adequately closeted and fringed by two full baths, and a handy laundry area is tucked behind the dining room.

**House—1,462 sq. ft., Garage—528 sq. ft.,
Breezeway—140 sq. ft.**

GARLINGHOUSE

ORDER YOUR NEW
From a compar

- Experience the thrill of creating and customizing your home.
- Enjoy comparing and choosing options.
- Obtain construction bids . . . and you're ready to build.

For 75 years, Garlinghouse homes have been built by tens of thousands of families across the nation. The construction blueprints are complete . . . accurate . . . and contain all the information a builder needs to begin construction.

GARLINGHOUSE PLANS SAVE YOU MONEY

The costs of designing our homes are spread over a number of plan buyers, nationwide. Therefore, you pay only a fraction of what you would spend to have a home designed (specifically for you).

BLUEPRINT MODIFICATIONS

It is expensive to have plans modified by professional designers. However, minor alterations in design, as well as building material substitutions, can be made by any competent builder according to the needs or wishes of the home owners.

ANOTHER GARLINGHOUSE DESIGN

GARLINGHOUSE PLAN SERVICE
No 9950

YOU CAN CHARGE YOUR ORDER!

TO ORDER PRODUCT LITERATURE SEE PAGES 95-96

BLUEPRINT ORDER FORM

PLEASE SEND ME:
- ☐ One Complete Set of Blueprints
- ☐ Minimum Construction Package five sets
- ☐ Standard Construction Package eight sets

Plan Number _____ ☐ as shown   ☐ reversed

Cost ....$_____

_____ **Additional Set(s)** $15.00 each........$_____

Materials List ($10.00 per order) ....$_____

Mailing Charges ....$___4.25____

TOTAL AMOUNT ENCLOSED.............$_____
(Kansas residents add 3 1/2%)

CHARGE MY ORDER TO:
- ☐ Mastercharge   ☐ Visa   ☐ American Express
  Exp.
Card # _____ Date _____

Signature _____

WE WOULD APPRECIATE YOUR HELP IN ANSWERING THE FOLLOWING QUESTIONS:

	Yes	No
Do you now own the land you plan to build on?	☐	☐
Do you plan to start construction in the next 6 months?	☐	☐
Do you plan to build most of the home yourself?	☐	☐
Would you like for us to have free building product information sent to you?	☐	☐

Name _____

Address _____

City_____ State_____ Zip_____

The Garlinghouse Co., 320 S.W. 33rd St., P.O. Box 299
(913) 267-2490   Topeka, Kansas 66601-0299

BLUEPRINT ORDER FORM

PLEASE SEND ME:
- ☐ One Complete Set of Blueprints
- ☐ Minimum Construction Package five sets
- ☐ Standard Construction Package eight sets

Plan Number _____ ☐ as shown   ☐ reversed

Cost ....$_____

_____ **Additional Set(s)** $15.00 each........$_____

Materials List ($10.00 per order) ....$_____

Mailing Charges ....$___4.25____

TOTAL AMOUNT ENCLOSED.............$_____
(Kansas residents add 3 1/2%)

CHARGE MY ORDER TO:
- ☐ Mastercharge   ☐ Visa   ☐ American Express
  Exp.
Card # _____ Date _____

Signature _____

WE WOULD APPRECIATE YOUR HELP IN ANSWERING THE FOLLOWING QUESTIONS:

	Yes	No
Do you now own the land you plan to build on?	☐	☐
Do you plan to start construction in the next 6 months?	☐	☐
Do you plan to build most of the home yourself?	☐	☐
Would you like for us to have free building product information sent to you?	☐	☐

Name _____

Address _____

City_____ State_____ Zip_____

The Garlinghouse Co., 320 S.W. 33rd St., P.O. Box 299
(913) 267-2490   Topeka, Kansas 66601-0299